Ecological Studies

Analysis and Synthesis

Volume 56

Ecological Studies

Volume 46
Eutrophication and Land Use
By William M. Lewis, Jr., James F. Saunders III, David W. Crumpacker, and Charles Brendecke
1984. X, 202p. 67 figures, cloth
ISBN 0-387-90961-3

Volume 47
Pollutants in Porous Media
The Unsaturated Zone Between Soil Surface and Groundwater
Edited by B. Yaron, G. Dagan, and J. Goldshmid
1984. XVI, 296p., 119 figures. cloth
ISBN 0-387-13179-5

Volume 48
Ecological Effects of Fire in South African Ecosystems
Edited by P. de V. Booysen and N.M. Tainton
1984. XVI, 426p. 54 figures. cloth
ISBN 0-387-13501-4

Volume 49
Forest Ecosystems in Industrial Regions
Studies on the Cycling of Energy, Nutrients and Pollutants in the Niepolomice Forest, Southern Poland
Edited by W. Grodziński, J. Weiner, and P.F. Maycock
1984. XVIII, 277p. 116 figures. cloth
ISBN 0-387-13498-0

Volume 50
The Gulf of Agaba (Elat)
Ecological Micropaleontology
By Z. Reiss and L. Hottinger
1984. VIII, 360p., 207 figures. cloth
ISBN 0-387-13486-7

Volume 51
Soil Salinity under Irrigation
Processes and Management
Edited by I. Shainberg and J. Shalhevet
1984. X, 358p., 133 figures. cloth
ISBN 0-387-13565-0

Volume 52
Air Pollution by Photochemical Oxidants
Formation, Transport, Control and Effects on Plants
Edited by Robert Guderian
1985. XI, 346p., 54 figures. cloth
ISBN 0-387-13966-4

Volume 53
The Gavish Sabkha
A Model of a Hypersaline Ecosystem
Edited by G.M. Friedman and W. E. Krumbein
1985. X, 484p., 246 figures. cloth
ISBN 0-387-15245-8

Volume 54
Tidal Flat Ecology
An Experimental Approach to Species Interactions
By Karsten Reise
1985. X, 198p., 69 figures. cloth
ISBN 0-387-15447-7

Volume 55
A Eutrophic Lake
Lake Mendota, Wisconsin
By Thomas D. Brock
1985. XII, 308p., 82 figures. cloth
ISBN 0-387-96184-4

Volume 56
Resources and Society
A Systems Ecology Study of the Island of Gotland, Sweden
By James J. Zucchetto and Ann-Mari Jansson
1985. X, 256p., 70 figures. cloth
ISBN 0-387-96151-8

James Zucchetto
Ann-Mari Jansson

Resources and Society

A Systems Ecology Study of the Island of Gotland, Sweden

With 70 Figures

Springer-Verlag
New York Berlin Heidelberg Tokyo

James Zucchetto
National Research Council
Energy Engineering Board
Washington, D.C. 20418
U.S.A.

Ann-Mari Jansson
University of Stockholm
Institute of Marine Ecology
Askö Laboratory
10691 Stockholm, Sweden

Library of Congress Cataloging in Publication Data
Zucchetto, J.
Resources and Society.
(Ecological studies; v.56)
Bibliography: p.
Includes index.
1. Natural resources—Sweden—Gotland.
2. Gotland (Sweden)—Economic conditions.
3. Power resources—Sweden—Gotland.
4. Ecology—Mathematical models. I. Jansson, A.-M.
II. Title. III. Series.
HC373.5.Z83 1985 333.7′09486 85-9955

ISBN-13: 978-1-4612-9574-7 e-ISBN-13: 978-1-4612-5124-8
DOI: 10.1007/978-1-4612-5124-8

Softcover reprint of the hardcover 1st edition 1985

Typeset by David E. Seham Associates Inc., Metuchen, New Jersey.

9 8 7 6 5 4 3 2 1

Preface

Although this book is about a specific area of the world (i.e., Gotland, Sweden), the interdisciplinary nature of the study, with regard to resources, environment, and society, makes it of interest to a number of fields. We have tried to make this book readable for a wide variety of interested parties including systems ecologists, environmental scientists, resource economists, geographers, regional planners, and regional scientists, as well as those interested in Nordic conditions. Since this project was part of UNESCO's Man and the Biosphere (MAB) program, this book should be of general interest to the international community. This book is certainly not a textbook, but we see it as being useful for courses in regional analysis with plenty of examples for illustrating analysis and models related to energy, environment, and economics, or to the general field of systems ecology. An instructor could, of course, supplement the material on systems and models with other sources. We hope this small book will serve as a helpful example of the analysis of the complex interdisciplinary problems associated with resources and society.

In Chapter 1, we present a brief introduction to the Gotland study as well as to some of the concepts and theories that have guided our investigations. Chapter 2 contains a detailed description of Gotland, as well as the background of our project. Both the historic and existing situation is presented, with regard to the economy, resources, and environment. It is not only descriptive, but also contains the empiric results of many energy, economic, and resource analyses that were conducted for different sectors and activities on the island. Part of the results from a previous monograph (Jansson and Zucchetto 1978a) are also succinctly summarized in places. Chapter 3 is devoted to the mathematic modeling undertaken for the economic, energy, and ecologic systems. Chapter 4 highlights the major results and findings of the project and their implications for Gotland. Finally, in Chapter 5, we attempt to draw some general lessons about systems ecology and our project for studies of humans and their environment. Comparisons are made to other studies, theoretic concepts are considered, and shortcomings and directions for future research are addressed.

Note on Units

Joules (J) have been used as the basic energy unit with prefixes as follows: k (thousand), M (million), G (billion), T (trillion), P (quadrillion). Electricity has been expressed in kWh. Swedish money is Swedish crowns or Swedish kronor (Skr). In 1978, 1 U.S. dollar was approximately 4.60 Skr. In 1985, the exchange rate is about 9 Skr to the dollar. Some useful conversion factors are as follows:

1 Kcal = 4.187 kJ
1 kWh = 3600 kJ
1 Kcal = 3.968 BTU
Heating Oil 1: 35.59 GJ m^{-3}
Heating Oil 3–5: 38.94 GJ m^{-3}
Diesel: 35.59 GJ m^{-3}
Kerosene: 34.96 GJ m^{-3}
Gasoline: 31.4 GJ m^{-3}

Acknowledgments

The Gotland project and this book would not have been possible without the assistance of many individuals and institutions. Funding has been provided through grants from the Swedish Natural Science Research Council, the Swedish Energy Research and Development Commission, the Swedish Council for Planning and Coordination of Research, fellowships from the Rockefeller Foundation and the University of Pennsylvania, and support for exchange students from the Swedish Institute. Those who spent considerable time on the project and whose work has been drawn upon are Helen Ahlbom, Ing-Marie Andréasson, Kazuo Furugane, Tuija Hilding, Karin Limburg, Torbjörn Nilsson, Maj-Liz Nordberg, Hideyo Shimazu, and Gary Spiller. Other contributions were made by Dan Carlsson, Lars Emmelin, Anders Engquist, Carl Folke, Stefan Häger, Helena Nantin, Göran Östblom, Christian Schaar, Bo Sundström, Elizabeth Titus, and Cristina Thoms. During the later part of our project, a reference group of advisors consisted of Erik Berggren, Thomas B. Johansson, Karl Göran Mäler, Bo Lindborg, Bengt Lundholm, Gunnar Törnquist, Torgny Schütt, and Uno Svedin. In addition, Bert Bolin, Alf Carling, Erik Eriksson, Tomas Rosswall, and Leif Wastensson gave valuable scientific advice. Several knowledgeable people on Gotland taught us about the local situation: Lars Danielson, Folke Gustafsson, Bjorn Nilsson, and Eric Thiery, among others. We are especially grateful to Frank Golley for reviewing the manuscript and making valuable suggestions for its completion. The general inspiration supplied by the following sustained us through this endeavor: UNESCO Man and the Biosphere program, especially John Celesia and participants in meetings of MAB-11 projects on urban ecosystems, colleagues from the Askö Laboratory at the University of Stockholm, the Regional Science Department at the University of Pennsylvania, and the Department of Environmental Engineering Sciences at the University of Florida. Our special thanks go to Howard T. Odum for his contributions to systems ecology, from which we have derived much of our inspiration. The artistic drafting was done by Bibbi Mayrhofer and most of the word-processing was done by Kathie Klingler. Last, but not least, we thank our families for enduring our obsession and preoccupation.

Contents

1. Introduction, Theory, and Perspectives

The 1970s was a time in which the era of explosive economic growth since the end of World War II was slowing, if not ending. Development came under serious analysis, criticism, and questioning in many nations. Was it possible to have ever increasing levels of economic wealth? What was happening to our basic life support system of the biosphere as industrial activity and personal consumption increased? Were there enough resources and energy to sustain a much longer period of rapid economic growth? People began to look for an alternative paradigm or view of the world that would be better suited to developing a sustainable system of humans and nature in contradistinction to much of the short-term economic gain that so strongly influenced corporate and governmental decision making. The environmental movement of this period gained ground as the realization of the deterioration of the environment became abundantly manifest and as the connectivity of the biosphere became a predominant theme (Ehrlich and Ehrlich 1970, Commoner 1971, Meadows et al. 1972, Watt 1974). If this were not enough ammunition to question the goal of continued economic growth, the oil embargo of 1973 was a dramatic illustration of the dependence of the industrialized nations on energy and of the finite and limited storages of fossil fuels, which were the primary energy sources. Political, public, and scientific interest in problems associated with environment, resources, and economics grew rapidly and many studies, programs, laws, and scientific theories emerged. One school of thought relating to these issues, which was promulgated by the well-known systems ecologist Howard T. Odum (1971, 1973,

1983), sought ecologic and energy principles as unifying factors for both ecologic and human systems. The emphasis on system properties and energy principles and measures inherent to this school of thought inspired and influenced our study of the island of Gotland. It is also the underlying theme of this entire book.

The influence and success of a Cartesian, mechanistic view of the world that promotes reductionism and the isolation of ever smaller units of the natural world to generate explicit relationships could not be denied. Numerous achievements in the physical and biological sciences illustrate the power of a reductionistic approach. However, it was this success that seemed to be generating such significant environmental consequences. A disturbing undercurrent arose in the face of all this success that was denoted by such terms as "holism" or "system views of the world" (Von Bertalanffy 1968, Laszlo 1972, Pattee 1973). In terms of environmental problems, it was imperative to understand that human action affected the biosphere, which in turn affected the human system; a systems view of humans and nature was needed. Systems concepts and analysis have been a natural part of the field of ecology for many years because of the emphasis directed towards understanding whole ecosystems, the relationships among plants, animals, and their environment, and the search for overall ecosystem properties that apply at the macroscale and that are generalizable across systems (Watt 1968, Van Dyne 1969, E.P. Odum 1971, Patten 1971–1976, H.T. Odum 1983). The progression of ecology to systems ecology, which included the systems of humans, gave rise to a new scientific field for dealing with the interactions between humans and nature (Holling 1978, Watt 1982, Odum 1983).

In systems ecology, the emerging problems related to resource use, environment, and economics are seen as highly interrelated. The energy disruptions that occurred in the 1970s and the ensuing economic malaise are thought to display the intimate connection between energy and economics. In addition, the environmental disturbances associated with economic activity and energy consumption have increasingly required resources to be dealt with. Many research endeavors were undertaken to synthesize the seemingly disparate disciplines of ecology and economics. One of the aims of the Gotland study was to examine working methods for such an integration (see Chapter 3).

The study and management of the relationships of natural resources to economics has been considered at several spatial levels of scale, from the global to the neighborhood, but much interest lies in intermediate or "regional" levels. Although the definition of the boundaries of a region is somewhat arbitrary, it refers in the present context to some extensive geographic space entailing a mosaic of ecosystems and human settlements that interact among themselves and with the physical environment. The study of Gotland was initiated in 1975. It was inspired by several studies that had been conducted for the region of South Florida in the first half of the 1970s and that the present authors had connections with. (Zucchetto, 1975a,1975b, Browder et al. 1976). As discussed in Chapters 2 and 3, the Gotland project focused most of its attention on the quantitative evaluation of material, energy, and economic flows in the region. The project was not designed primarily for the solution of specific planning problems, although the approaches and results, suitably modified, could be conceivably incorporated into different aspects of regional planning. Although we were well aware of the potential social consequences of changes in the economy, environment, or energy supply system, the scope of the Gotland project was not extensive enough to include any meaningful explorations of the social system. Because of extensive income equalization policies, impacts on disparities among different social groups was not as important a consideration as it might have been in other nations.

The study of island ecosystems constitutes an important area of interest in the field of ecology (MacArthur and Wilson 1967). Although our focus was not on biogeography, the study of economic/ecological interactions is of particular interest for the understanding of island development (Brookfield 1980, see also Chapter 5, the section on "The Case of Eastern Fiji"). Due to their well-defined boundaries and restricted size, islands usually are easily modeled and can be used as pilot studies to test out models of larger, more complex mainland areas. Island systems are subject to the vagaries of external forces, whether they are economic conditions in the world market or natural dynamics of the biosphere such as weather. Their isolation and size are related to the vulnerability of their environment. Exploitation of resources and population increases can generate severe impacts requiring careful environmental management. In particular, the coastal zone is important to the total island system and is subject to impacts from surrounding seas. However, the isolation of islands is also, to some extent, protective; it produces unique conditions that in many instances enhance their image and charm for foreigners. For Gotland, the relationship to mainland Sweden via transportation, energy supply, and economic transfer payments has greatly influenced its recent development and altered much of its previous character.

Throughout the study period, the following goals and objectives were defined and, to one extent or another, reasonably met:

1. To demonstrate the use of a systems ecology approach to regional systems of humans and nature that would engender mutual interaction among members of different disciplines. Gotland was used as a representative study site for demonstrating the approach of a systems ecology methodology to an existing system of humans and nature. Interested parties in Sweden and elsewhere were made aware of this work through meetings, seminars, and many publications. In 1977, the Gotland project became a part of the UNESCO Man and Biosphere (MAB-11) program, with consequent dissemination of information and interaction with other MAB participants (Jansson and Zucchetto 1980).
2. To quantify large-scale measures that would be applicable to the total regional system, and to monitor these over time.
3. To conduct detailed energy-economic analysis of economic system activities and energy analysis for the natural systems.
4. To identify and quantify the couplings between the human system and the natural systems. These couplings might take the form of harvesting or pollutant generation by the human system.

5. To formulate mathematic simulation models for the economic and natural systems to predict impacts of various changes in activities.
6. To formulate optimization models to demonstrate the notion of trade-offs between activities subject to various kinds of constraints.
7. To investigate the potential contribution of renewable energy technologies to the region.
8. To apply both the systems analysis technique and remote sensing technology to the understanding of land use and associated environmental repercussions in a subarea of Gotland. This was accomplished in the area of Lummelunda, north of the capital city, Visby.
9. To accumulate information and formulate models that could serve as an educational introduction to the holistic view for the planners of Gotland, as well as to describe a repertoire of techniques that could be incorporated into the planning process.

In general, the project accomplished these goals to one degree or another.

Models and Systems

By the very nature of their minds, humans tend to organize their conception of the environment with the aid of some form of model representing an abstract simplification of the world. One of the basic approaches taken in this present study was to consider the region of Gotland as a complex system—a collection of units interacting with one another to produce an identifiable whole (Laszlo 1972). In this case, the "whole" is the island of Gotland itself, and the units may be economic sectors or ecosystems (Figure 1.1). Although some components may change, there is still some identifiable larger system. In general, the units that comprise the system may be systems themselves and so on, ad infinitum, thus comprising a system of systems. For example, a systems model representing the interactions of the main economic sectors of Gotland includes agriculture as one unit. On more detailed inspection, agriculture itself becomes composed of several interacting units. The definition of a system will depend very much on the view of the investigator and the questions of interest; once this is clarified a boundary can be chosen that defines the system. Everything within the system boundary is then regarded as part of the system, and everything external to the boundary is the surrounding environment that interacts with the system. The boundary of the system may be defined by spatial considerations (e.g., a watershed or the shoreline of an island) or functional considerations (e.g., all of those components related to a health care system in an urban area). For our study, the well-defined boundaries of an island facilitated the evaluation of inflows and outflows, as well as the identification of external influences operating on the local economy and ecology. The power of the systems approach is that it deals with complexity in an organized fashion, allowing one to determine the impact of one part on all of the others.

Several degrees of sophistication can be used to formalize the interaction

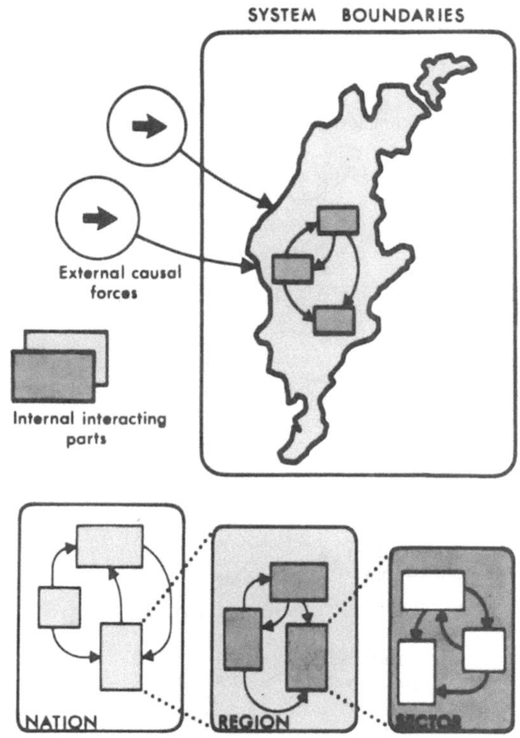

Figure 1.1. General diagram of a system shown with internal components interacting. These internal components are also systems, so a "system of systems" is conceptualized. The environment affects the system and the system affects the environment through transfers of energy, matter, and information. The system outline in this case is the island of Gotland.

among system components. At the qualitative level, mapping chains of cause and effect can help to identify and elucidate mutual dependencies. For example, Figure 1.2 maps out two chains of events that have figured importantly in the region of Gotland; i.e., the relationship of energy to the economy and water quality. These schemes need to be more precisely defined for the purpose of scientific investigation; in themselves, however, they can serve as an important mode of communication to the nontechnical investigators in a particular setting to integrate their knowledge about existing interactions.

More precise systems models incorporate actual measures of flows from one sector to another. For example, our initial work constituted the identification and evaluation of major storages and flows for several systems on Gotland. These were illustrated with systems diagrams using the energy flow language of H.T. Odum (1972). They included natural ecosystems and activities such as agriculture, forestry, fisheries, and the regional economy (Jansson and Zucchetto 1978a). These models served as a template for analyzing the use of energy and materials, as well as for guiding fieldwork and data collection. Certain models

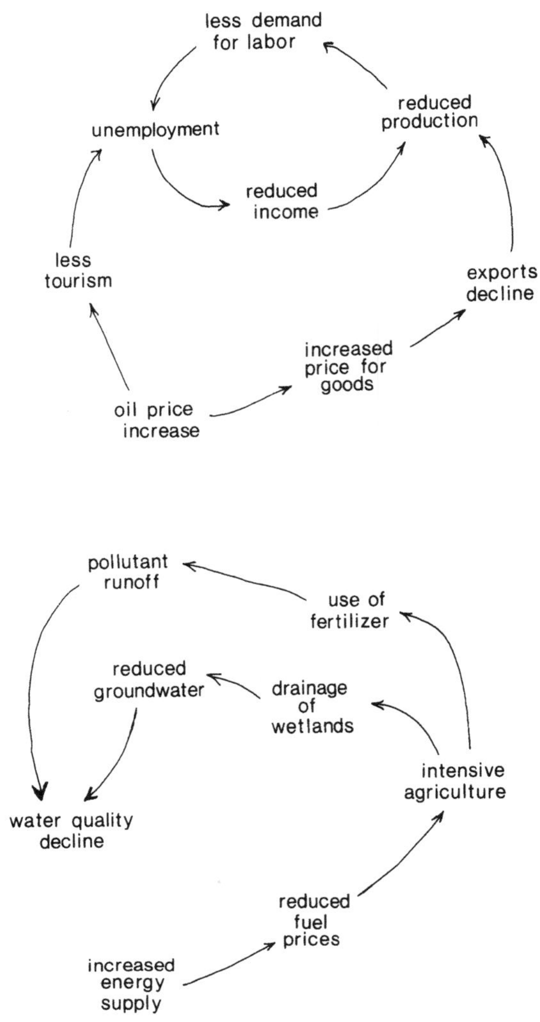

Figure 1.2. Qualitative examples of the chain of events that can be triggered in a complex regional system.

were translated into systems of linear or nonlinear differential equations for the study of system dynamics. For example, a model of the hydrologic system on Gotland was simulated to ascertain the impact of fertilization regimes on groundwater quality. Differential equation models were formulated for the coastal ecosystem with the aim of investigating the impact of nutrient runoff from land and fish harvesting.

A convenient representation of an economy is an input-output model that incorporates the input requirements from all sectors to a given sector, assuming that there is a linear relationship between the inputs and the output of a given sector. We used input-output analysis for the entire economy of Gotland to calculate the resource requirements and pollutant generation for different levels of economic output. These were also coupled with linear programming models, which maximized economic output, to ascertain the impact on economic activity of restricting pollutant emissions or available fresh water. An input-output model was also modified to simulate the effect of energy supply disruptions on the economic output of Gotland's economy. Finally, optimization models were formulated to determine potential contributions from renewable energy technologies.

Resource, Energy, and Environmental Analysis

Energy Analysis

From the very beginning of the Gotland project, the primary emphasis has been on a systematic energy analysis of both the human and natural systems on the island. From this point of view, energy served as a convenient thread weaving together the activities of both humans and nature. Understanding the relationship of energy to the functioning of economic systems is important to plan better for uncertainties in energy supply and price, as well as for increased efficiency of use. Knowledge of the energy fluxes associated with the natural environment helps to evaluate the work that nature performs, the yields that are available for harvest, and the potential contributions that are yet to be directly exploited by humans. In this energy view of a region (Figure 1.3), there are various sources

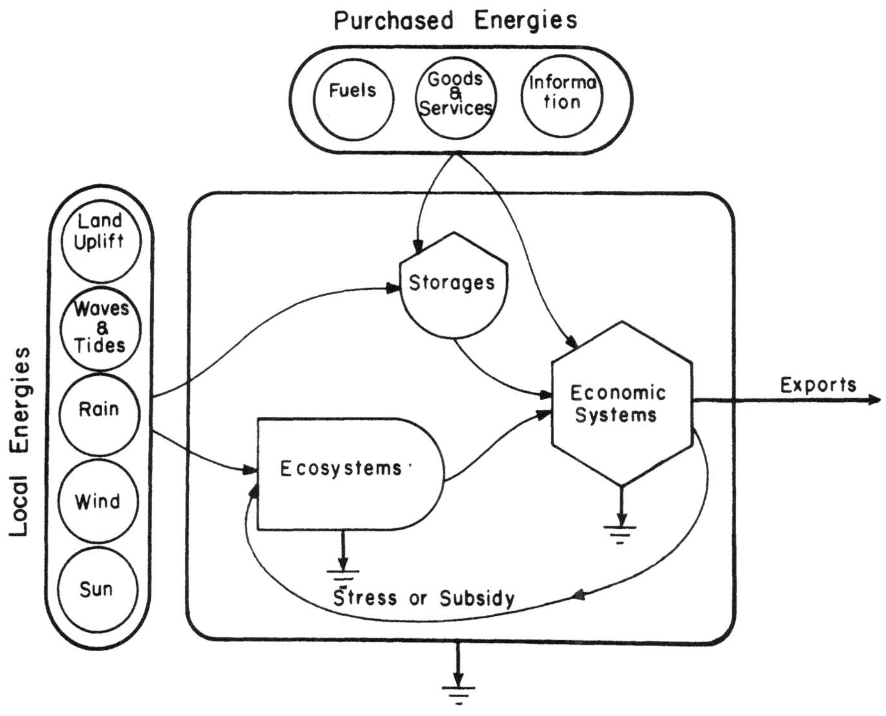

Figure 1.3. Generalized model of a region showing dependence on natural energies, imports, and nonrenewable resources. The economic system interacts with these, affects natural systems, and generates exports.

of indigenous energies, both renewable and nonrenewable, as well as purchased imports of energy, goods, and information both supporting activities in the economy and generating exports. Significant environmental changes occur that either enhance or decrease the productivity of the natural ecosystems.

The initial overview evaluation of the energy flows on Gotland included the spectrum of renewable flows of energy available to the island. The fixation of solar energy by photosynthetically based terrestrial and aquatic systems, such as forests, agroecosystems, wetlands, and the coastal waters was calculated from data on productivity and harvest. The use of imported fuels and electricity in these systems, as well as in a number of sectors of society, was also analyzed. In many instances, time series data were collected to assess trends. Imports and exports of goods were also estimated for the purposes of determining the energy cost, which was expended outside the region, for supplying these goods to Gotland's economy. These data not only formed a detailed quantitative energy picture of the island, they were required for parameterizing the various economic and ecologic models that were formulated during the course of the study.

Energy Quality

It is known that all forms of energy can be converted into heat. However, another fundamental characteristic is the ability of different forms of energy to do work, which is sometimes referred to as the quality of the energy source. There are several notions of energy quality that may shed light on this subject. Under controlled conditions, a more concentrated energy source can perform more work than a less concentrated source. Higher concentrations of energy in space yield greater temperatures, which produces a greater potential for useful work for a given amount of energy. Dilute energy sources, such as sunlight falling on the earth's surface, require collection and concentration before being able to do useful work. A scale of energy quality could be calculated in which a progressive concentration of energy is obtained by actions of the biosphere and humans, for example, from sunlight to fossil fuels to electricity (Odum 1973, Odum and Odum 1976). There is an upgrading in energy quality that occurs; it is paid for by a second law energy cost at each step. Second is the notion that quality refers to versatility, flexibility, and ease of use from technical as well as environmental points of view. Electricity, once it is produced, can be used in diverse ways (from driving mechanical systems to information processing) and is relatively benign environmentally. A final measure of quality can be determined by considering the energy required to perform a given task: if it takes fewer units of energy type 1 than type 2 to perform a given task, then type 1 can be thought of as having a higher quality than type 2. For example, if traversing a given distance in a certain time required 5 units of oil as opposed to 10 units of coal the quality of oil would be twice that of coal, for accomplishing the task.

Efficiency

The term efficiency generally connotes the ratio of output to input for a process. For an energy system such as a power plant, concern lies with the output of electricity per unit input of fuel; for an economic system, one might be concerned with some measure of economic output (in numbers, weight, or economic value) per unit of input of energy or labor, or other factors. Later in this chapter, ratios of energy to gross economic output are discussed as one measure of efficiency. In the Gotland study, the detailed energy analysis of various activities has allowed the calculation of energy efficiencies. For the natural and managed ecosystems, estimates of photosynthetic efficiencies, yield efficiencies, and energy costs per unit of output have been made. For the human systems, time series data for industry sectors of energy per unit of economic output have been calculated (see also the section on "Energy and Money"), as well as other efficiency measures in transportation and energy production. In some cases, comparisons have been made to other studies.

Energy and Money

The events of the 1970s have pointed out, even to the most adamant observer, the extreme importance of energy for human activity and processes in human economies. Sources of inanimate energy, in addition to the flows of energy in the biosphere, are inherent in all present-day activity (Fairgrieve 1921, Soddy 1935, Cottrell 1955, Lotka 1956, Hubbert 1962, Odum 1971, Cook 1976, Slesser 1978). The productive processes of a national economy can be measured in terms of the gross national product (GNP), in monetary terms, of goods and services. Past studies have been directed towards identifying a relationship between economic output and energy consumption for either regions, individual economies, or across national economies (Barnett 1950, Gambel 1964, Cook 1971, Kylstra 1974, Zucchetto 1975a, 1975b, 1985a, 1985b, Darmstadter et al. 1977, Hall et al. 1984). Typically, a correlation can be observed between energy consumption and gross economic output for a given economy (Figure 1.4). A smaller slope would imply a lower efficiency. The ratio of energy to economic output is also interpreted as a measure of the inefficiency with which energy is used to generate economic output; a higher ratio implies lower energy efficiency. These are macromeasures of an economic system. Their variation with

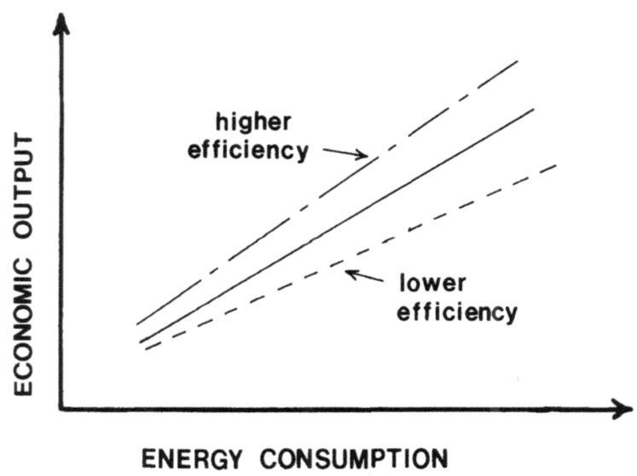

Figure 1.4. Typical graph of gross economic output versus energy consumption. Steeper slopes could be interpreted as higher energy efficiency.

time can serve as a comparative measure between systems or as an indicator of ongoing trends.

Energy consumption usually refers to the direct use of fuels and electricity in an economy, but some studies also consider energy flows in the natural systems. The linkage between the processed energies and economic activity is rather evident. Contributions to human economies from natural systems are more difficult to assess but thermodynamic completeness necessitates their inclusion. For example, if the use of natural resources entails the exhaustion of soil in a given area, then the energy cost of reestablishing that soil should, in fact, be included in an overall assessment of the energy-economic relationship. Continuous flows of renewable energy that provide goods and services should be considered as contributing to an economy.

Establishing the relationship between energy and economy activity, and its trend over time, is important information in regional planning. If this relationship is invariant over time, then anticipation of future levels of economic activity must be considered in light of available energy. Furthermore, energy conditions will dictate economic growth; scarcity will inhibit growth and abundance will stimulate it. If the energy-economic ratio is changing over time because of increased efficiency, then the degree to which it changes gives additional information on future potential for economic activity.

For Gotland, both the gross regional product (GRP) and energy consumption were evaluated over time to assess trends. Measures similar to the energy-to-GNP ratio were also defined for sectors of the economic system. For example, the measure analogous to GNP for industries is "value added," which is the amount of sales minus the cost of raw materials that the industry requires. The total annual energy consumption in that industry divided by the value added could be used as a measure of energy intensiveness or, again, as an efficiency measure. This measure varied substantially across industries. Energy-economic measures were also established for agriculture, fisheries, and forestry.

Embodied Energy and Energy Cost Calculations

The analysis of the structure and function of both natural and human systems requires the careful tracking of matter and energy throughout the system. Ecological energetics entails the determination of the extent to which solar energy is transformed to energy stored in biochemical compounds, the distribution of this stored energy throughout the ecosystem, and the ways in which energy flows affect the various biotic and abiotic components of the ecosystem (E.P. Odum 1971). Likewise, for a human economy, energy analysis focuses on the way in which solar energy and fuels contribute to various human activities and the consequences of this use. In both systems, there exists a complex web, or simplified, a chain of transfers of materials and energy that begins with primary, simple raw materials. Work is done on these materials, resulting in complex biochemical compounds in ecosystems or products and human services in human economies. This structure can be portrayed as shown in Figure 1.5 in which a simple food chain, a food web of an ecosystem, and the transfer between sectors

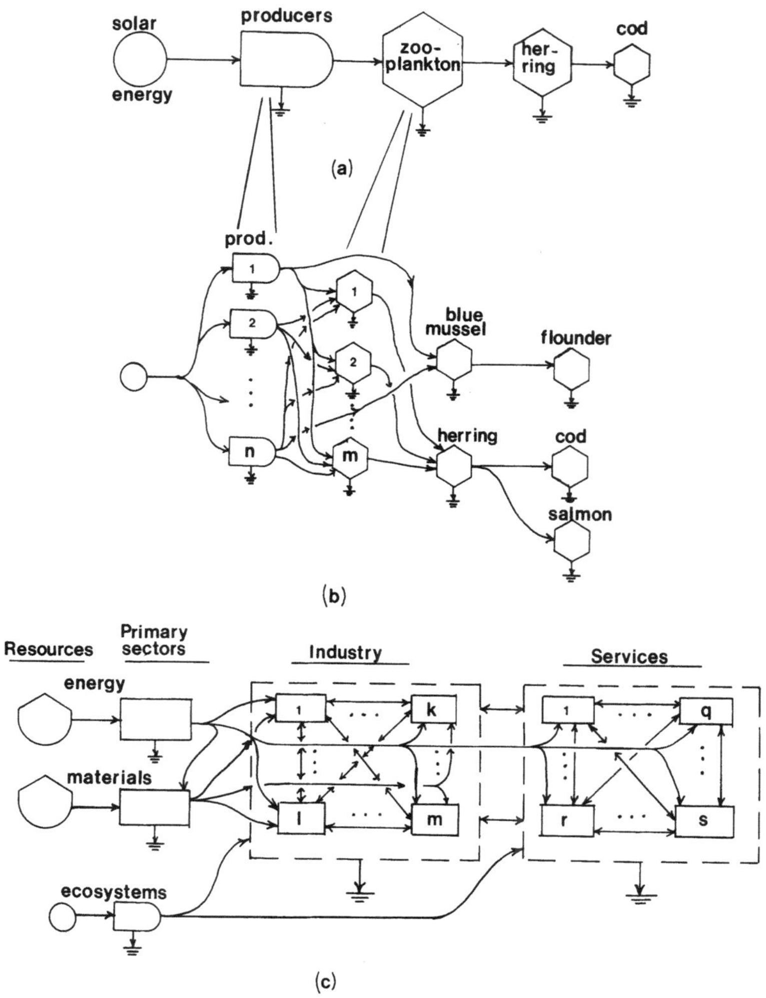

Figure 1.5. (a) Example of simple ecosystem food chain. (b) Food web in an ecosystem. (c) Web of activities in an economy.

in an industrial economy are compared. For the Gotland economy, sunlight supports complex and economically valuable members in the coastal ecosystem, such as salmon and mink, while imported electronic component parts are upgraded in the workshops industry, to functioning telecommunication equipment.

Embodied energy or energy cost is the energy required, both directly and indirectly, i.e., throughout the system per unit of output for a given component of the system. If 1000 J (joules) of solar energy produce 10 J of phytoplankton and 10 J of phytoplankton produce 1 J of zooplankton, then the solar energy embodied in zooplankton would be 1000 J/J. Similarly, for an economic system divided into sectors, each sector is generating output and is interacting with other sectors in trade. All of these flows are usually measured in dollars, but each sector requires energy (fuels and solar) to produce output. Thus, for the

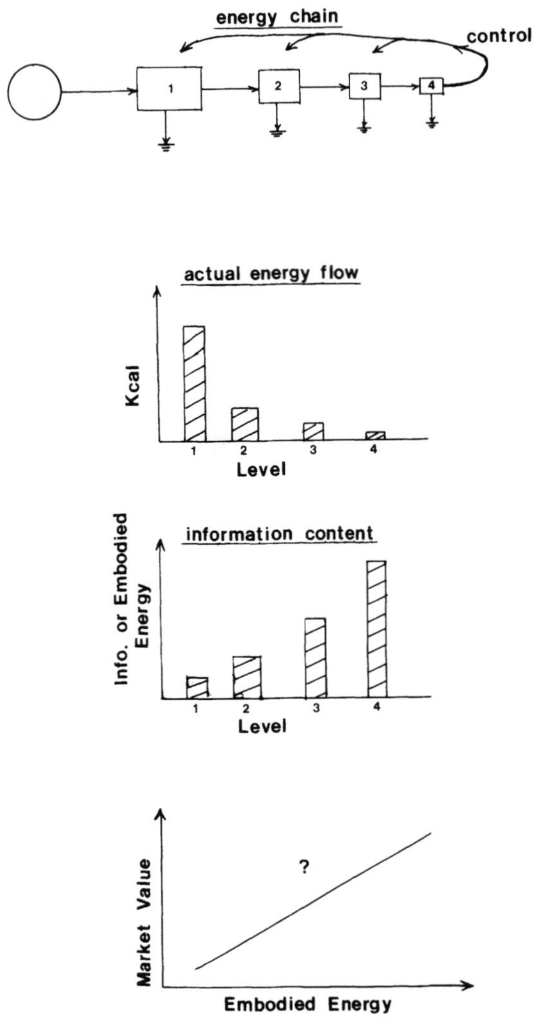

Figure 1.6. Notions of embodied energy. Components higher up the food chain have less actual energy flow, but more embodied energy and information content to exert control. Market value of products in an economy should be proportional to embodied energy.

cement industry on Gotland, energy is required to mine the limestone, to transport it to the factory, and to support and fuel the factory. Proceeding in this way, a calculation of the total energy (in energy units) throughout the economy for producing a unit of output (in money or physical units) from any given sector can be made. This would be the embodied energy or cost per unit of output (e.g., J dollar^{-1}). One might also consider the solar energy required to produce a resource such as limestone. This, in some sense, represents the replacement energy cost. Resource exploitation, such as limestone quarrying, usually entails the loss of productive ecosystems that also may be debited as an energy cost (see the section on "Net Energy Analysis"). Embodied energy ratios have been used in environmental evaluations to calculate energy costs associated with various projects and policies for comparative purposes (Kemp et al. 1977). Such energy cost assessments constitute one aspect for consideration in the decision-making process.

Components higher up the food chain or food web are more complex; e.g., top carnivores in ecosystems and sophisticated technology and human services in economies. These components, which are further removed from the initial energy sources, generally have higher embodied energy coefficients. In general, the higher up the food chain, the more complex the biochemical structure, the more intricate the behavior, and the greater the control of the environment. The higher trophic levels have lower total energy flows because of the second law losses all along the food chain resulting in a counterdistribution of information and energy (Figure 1.6). It is hypothesized that these complex and more highly ordered structures can amplify and control energy processes in the total system to facilitate the capture and use of additional energy (Odum and Odum 1976). If this is true, then in the economic system, this might imply a correlation between embodied energy of products and market value; high market value results due to their amplifier effect in the economy (see Figure 1.6c; see also Costanza 1980). On Gotland, for example, we see the high economic value associated with top carnivores such as salmon and mink. Additionally, complex components higher up the economic chain, such as telecommunication equipment and human services (both having high embodied energy), are also highly valued and are an important part of the development of the economy.

Net Energy Analysis

Energy cost analysis applied to the specific situation of energy technologies has led to the interesting concept of net energy. Although formalized quantitatively over the past 10 years by several investigators, the concept of surplus energy and its importance to the development of human systems was already introduced by Cottrell in the mid-1950s. It has been elaborated on in many studies (Cottrell 1955, Odum 1973, Chapman 1974, Gilliland 1975, 1978, Thomas 1977). In rather concrete terms, one can think of a simple hunting society or a population of predatory animals. The members of these two groups must expend energy in their hunting to find and capture food; if they expend more energy than they capture, then hunger, stress, and decline will result. If more energy is captured than expended, then growth and development can take place in the form of an increased population or enhanced specialization within the group. Depending on the abundance of food in the surrounding territory, there will be a distance beyond which net energy will not be gained; this will define the territory. Similarly, for modern societies, if the sources of coal, oil, wood, and other energies are far from points of use, deep in the earth, and of low quality, then it may take more energy to mine, process, and transport than is yielded at the point of use. Processing activities require goods and fuels that are supplied from the main economy; the energy cost of the energy processing or transformation activity includes direct energy consumed in the process, indirect energies consumed in the production of goods in the economy, and (in addition) natural system work activities that have been reduced or eliminated because of the energy processing activity. The useful energy produced minus the total energy costs, all of which are expressed in units of equivalent energy quality, is defined as the net energy of the source ($J_{Net} = J_{Out} - J_{In}$) while the ratio of output to cost is referred to as the energy yield ratio ($EYR = J_{Out}/J_{In}$) (Odum et al. 1976). Calculations of net energy and energy yield ratios were carried through for several proposed, renewable energy technologies for Gotland. If it turns out that a proposed technology consumes more energy than it produces, then it would be unwise to invest in it, because it would have to be supported by other energy sources.

Resource and Environmental Analysis

Although our primary focus from the beginning of the study was on energy, it soon became evident that other resources were needed to be studied in some detail. Our interest focused on those resources and environmental issues that seemed most critical and limiting for economic activity. Water turned out to be the most crucial factor requiring study. With regards to water, attempts were made to estimate surface and groundwater supplies, as well as the flows and use of water in natural and human systems. These data were used in formulating hydrology models and in defining constraints on water use in the economic optimization models. They were also used in calculating multipliers, or total use of water, per unit of output from each economic sector. Water quality was also investigated to ascertain the impact of fertilization schemes on groundwater nitrate concentrations. This information proved useful in estimating constraints in the linear programming models. This was also true for data collected on pollutant generation of oxides of sulfur and nitrogen and on biological oxygen demand from different economic sectors. (Some analyses using air pollution models were conducted for the town of Slite with regard to point source emissions of sulfur dioxides. For now, this is not an important area of concern. Instead, emissions from oil-fired power plants, large ships, and traffic congestion in Visby have caused air pollution problems that should be addressed in future planning.)

One overall concern that was inherent in our project was the management of ecosystems for conservation or harvest. The modeling of the coastal ecosystem was set up to study both the influence of inputs of nutrient runoff from land and the impact of fish harvesting. Our calculations of forest productivity and exploitation of forests also could be used to estimate conditions for limits on wood harvesting. Another activity incorporating both economic and ecologic considerations is the exploitation of the limestone deposits on Gotland, which is resulting in a serious disruption of land, an elimination of a fragile ecosystem, and hydrologic impacts while generating large economic benefits.

In general, in any regional study, resource use must be studied in terms of its benefits to a society as well as its impact on the environment and ecology. To this end, quantitative and qualitative data must be gathered and models must be formulated, where appropriate, to help in the evaluation of economic and ecological impacts. Issues to be addressed should be determined by their perceived importance and should not be hampered by insufficient resources available to conduct the analysis.

Ecologic Perspectives

Ecology is the study of the interrelationships among organisms or groups of organisms and their interactions with a given environment (E.P. Odum 1971, Ricklefs 1973, Watt 1973). The term "organisms" rightly includes humans in the field of ecology. In a broad sense, systems ecology can be interpreted as the study of humans and nature. The field of human ecology deals to a greater extent with social and institutional organization. One of the primary interests of systems ecology is quantifing energy transfers throughout an ecosystem, as well as the cycles of materials generated as a result of the flow of energy through food chains and webs and in the biosphere as a whole. In our analysis, gross and net photosynthetic productivities were estimated for each of the major ecosystems, and attempts were made to determine the use of this production in the ecosystems as well as in the human system. The associated flows of nutrients, water, and organic matter were also evaluated in some situations. The dynamics of some of the ecosystems were investigated through the use of systems ecology models, as well as via the use of time series data. No specific studies on ecologic succession were undertaken, although some historic analysis of land use (over centuries and decades) indicated how the overall distribution of ecosystems in the landscape had changed.

Regional Systems and Ecologic Principles

General principles derived from the observation of ecologic systems may also lend some insight into human system organization and its interaction with the biosphere. Let us briefly explore some ecosystem properties that may have some relevance for the structure and function of human systems.

Succession

Do economic systems or regional systems obey some form of successional growth in which more and more of the productive work of the system would be channeled into maintenance and replacement, thus leading to an eventual dynamic steady-state, the climax? In this process, would we observe greater niche specialization (professions), more intricate food webs (economic transfers), greater information per unit mass, increasing diversity and closed nutrient cycles (material reuse and recycling), as suggested by E.P. Odum (1971) for natural ecosystems? Is the rapid growth of employment in the public sector for Gotland an indication of greater resources devoted to the maintenance and control of the economic system (see Chapters 2 and 4)?

Carrying Capacity

The notion of carrying capacity is commonly found in the ecologic literature, and it refers to the level of population that a given habitat can support. For a natural ecosystem, this carrying capacity is a function of the natural productivity in a given area. For a human system, it is more complex because of trade that allows a given human settlement to be supported by a vastly larger "hinterland."

However, there are still some inherent limits to population density in a given habitable area. As economic systems and human population grow, there are greater demands on resources, stress on the environment, and the generation of all forms of pollution. In a pragmatic sense, for a given standard of living, how many people could a region support given the constraints on available resources, land, and the generation of pollutants? An overall measure may be to find an optimum ratio of natural system energies to economic system energies (see Chapter 5).

Diversity and Stability

Although subject to debate, there is a body of literature suggesting that diverse ecosystems are resistant to changes in the environment. If the relationship between diversity and stability holds, then in the interests of the long-term viability of a region, policies should be formulated that encourage a diverse-mosaic of human and natural systems. Furthermore, the notion of stability should be tied to the availability of storages in a system that can be used during times of external interruptions. The existence of resource storages acts as a stabilizing influence during disruptions of external supplies. For Gotland, the strategy of diversifying energy sources was explored through simulation and input-output models (see Chapter 3).

Organization for Maximum Power

We address the maximum power hypothesis in Chapters 4 and 5 (Odum 1983). In relation to this controversial hypothesis, a number of intriguing questions arise. If systems are organizing for maximum power, what are the ways in which human systems tend to develop to maximize power flows? Could urban agglomerations (e.g., cities, towns) be regarded as means for accelerating energy flows of concentrated fossil and nuclear fuels? Is the scheme of increasing energy quality up through food chains—from sunlight to crops in rural areas to small towns to large cities, culminating in the flexible and complex modern societies—a means of exercising and orchestrating feedback controls on the total system of humans and nature to maximize power flows through the system? Can criteria of maximum power be used for choosing among a set of alternative land use arrangements?

Economic Considerations

The main economic concerns that we have dealt with in the Gotland project are of general scientific interest for most regional studies. Regional economics is a well established field (Isard 1975, Richardson 1978), but the methods are relatively little used in planning. At the macrolevel, the gross regional product (GRP, which is the total output of goods and services) is of interest. It may be desirable to increase the GRP in order to increase the material well-being of the population. Such goals may be accomplished in a number of ways by expanding different sectors in the economy. Which sectors are "best" to expand, given certain constraints on resources? For example, for Gotland, is it better to expand cement production or the food processing industry to expand total GRP? What are the employment, resource, and ecologic ramifications for expansion of these two activities? Time series data give some insight into trends in GRP and into why observed changes have occurred. Gross measures of energy and resource consumption, pollutant generation, and land use change give some overall impression of activity in a given region. Also of interest is the total energy use required per unit of GRP. As discussed previously, this can serve as a measure of efficiency or to help predict future levels of economic output in light of given constraints on energy availability. Most regional systems are open and dependent on trade. The Gotland economy's dependence on energy imports led to the consideration of what is the response to energy shortages and price increases, which can be generalized to include all critical imports. Furthermore, Gotland, as with most regional economies, depends on export markets that must be analyzed if predictions are to be made of the future level and kind of economic activity. In general, then, the next larger national and international economies exert the most important influence; they must be considered before local development can be predicted.

At a more microlevel of scale, such as at the level of a sector or a firm, information of interest is the consumption of resources, the economic activity generated, and the environmental repercussions produced. The Gotland study focused on energy and water as critical resources and levels of nitrates in groundwater as a critical environmental problem. Analysis of different economic activities produced quantitative measures of energy and water consumption, value added, salaries, labor, and pollutants produced; it allowed trade-offs among energy, labor, environment, and pollutants to be quantified. For example, energy consumption (in energy units) per unit of value added (in monetary units) is a measure of energy intensiveness for different activities. Suppose there is only a restricted amount of energy available for consumption. This information could be used, for example, to determine the distribution of energy among activities to maximize value added for the total region. Understanding the economic interactions among the various activities in a region also helps to determine the impact of changes in the local economy in either resource inputs or external markets. The trade-offs between economic activity and environmental impact can also be analyzed. If there is a change in one sector in the availability or cost of imports, or changes in the demand for exports, then a cascade of changes will be produced throughout the regional economy. The various resource, economic, and environmental impacts caused by these changes can be assessed with the help of regional systems models. An economic issue receiving much attention is employment. This has been an important consideration, since energy intensive agriculture in the post-World War II period has displaced human labor in Gotland's rural areas. Lacking employment opportunities on Gotland, many of these people have migrated to the mainland. In response, incentives have been created to encourage industry and government to locate on Gotland and to generate employment opportunities. The employment issue still remains important; certainly, one priority of an energy study is to consider the trade-off between energy and labor, which is a result of investing in either capital- or labor-intensive activities.

Plate 1. Trullhalsar. An old cemetery in a pine forest (Photo, T. Hilding).

2. Overview and Empiric Results for Gotland

Introduction

At the very beginning of the project, a primary goal of the Gotland study was to piece together in a quantifiable way the intricate flows of energy associated with both human and natural systems. This emphasis was, perhaps, made more immediate in the middle of the 1970s following the 1973 oil embargo and the turbulence of the world energy picture in terms of oil disruptions, shortages, and price increases. However, this is not to say that the Gotland project was limited to an "energy" study; rather, the roots of our approach were derived from systems ecology, which is a framework that led naturally to an integrated representation of the region in terms of resource (with the emphasis on energy), economic, environmental, historic and cultural considerations. The unique approach of the study was its emphasis on the interaction of various aspects of the region and, in particular, the coupling that existed between human activity and the natural environment. Empiric testing of energy theories of development and organization of systems, such as those proposed by H.T. Odum, was also an important goal. Research focused on the island of Gotland in the Baltic Sea for several reasons. As mentioned in Chapter 1, island systems are of interest because of their special characteristics. Gotland was heavily dependent on external sources of energy and, therefore, could serve as a representative study site for other regions in similar situations. Being an island, it also had well defined boundaries, which facilitated the collection of inflow and outflow data.

It also had a manageable level of complexity that could be dealt with, considering the limited extent of our research project. Furthermore, notions of carrying capacity could possibly be dealt with in a rather well-defined way. Several environmental problems also were of interest, including diminishing soil layers, shortages of water, and pollution of groundwater. These problems were partly a result of the extensive drainage of wetlands that has occurred over the course of 100 years.

In the early years of the project, the main emphasis was on the total system perspective, with the formulation of models and the collection of data that would describe the total regional system behavior in the face of changes in the conditions (especially energy) of the external world. During this process, a detailed understanding of the whole region emerged as well as of several activities within it. More applied problems were considered during the project's latter years, and greater emphasis was placed on economic analysis and energy technologies. The site specific area of Lummelunda, 20 km north of Visby, was investigated regarding land-use change with the use of remote sensing technology; land-use and its relation to water quality were studied with the aid of models. The results of these efforts are not presented in detail in this volume, but we refer to this study area in some places (Nilsson 1982, Nordberg 1983).

Background and History

Gotland is an island some 3100 km^2 in area located in the Baltic Sea about 200 km southeast of Stockholm (Figure 2.1). The bedrock consists of limestone dating back to Silurian times (400 million years ago), when this area was close to the equator and solar energy contributed the organizing work for building the calcareous skeletons of marine organisms that accumulated on the sea bottom to form the deposits. The existing terrestrial systems on Gotland have been formed by geologic processes in more recent times, just prior to the arrival of humans on the island. The last Ice Age engendered the movement of ice to positions well south of Gotland, resulting in deposits of moraine, sand, and clay layers on top of the bedrock. In addition, the weight of the ice sheets on the land, not only of Gotland but of the mainland areas surrounding the Baltic Sea, compressed the lithosphere. Since the ice sheets receded, a return to equilibrium has resulted in the Baltic coast of Sweden having risen some 200 m.

Gotland is still rising at an average rate of about 0.25 cm yr^{-1}. The rate of rise of the island is somewhat larger in the north than in the south, with this phenomena leading to an increase in area of the island. If the level of the Baltic Sea were to remain constant, then this process of geologic uplift would result

Figure 2.1. Satellite photo of the island of Gotland, Sweden, with its geographic location in the Baltic. The island measures some 130 km in length and 50 km at its greatest width. Gray areas are mostly forest and white areas are mostly agricultural. Main towns and highways are indicated in black. (Landsat II, June 12, 1975, MSS 7) (Jansson and Zucchetto 1978a).

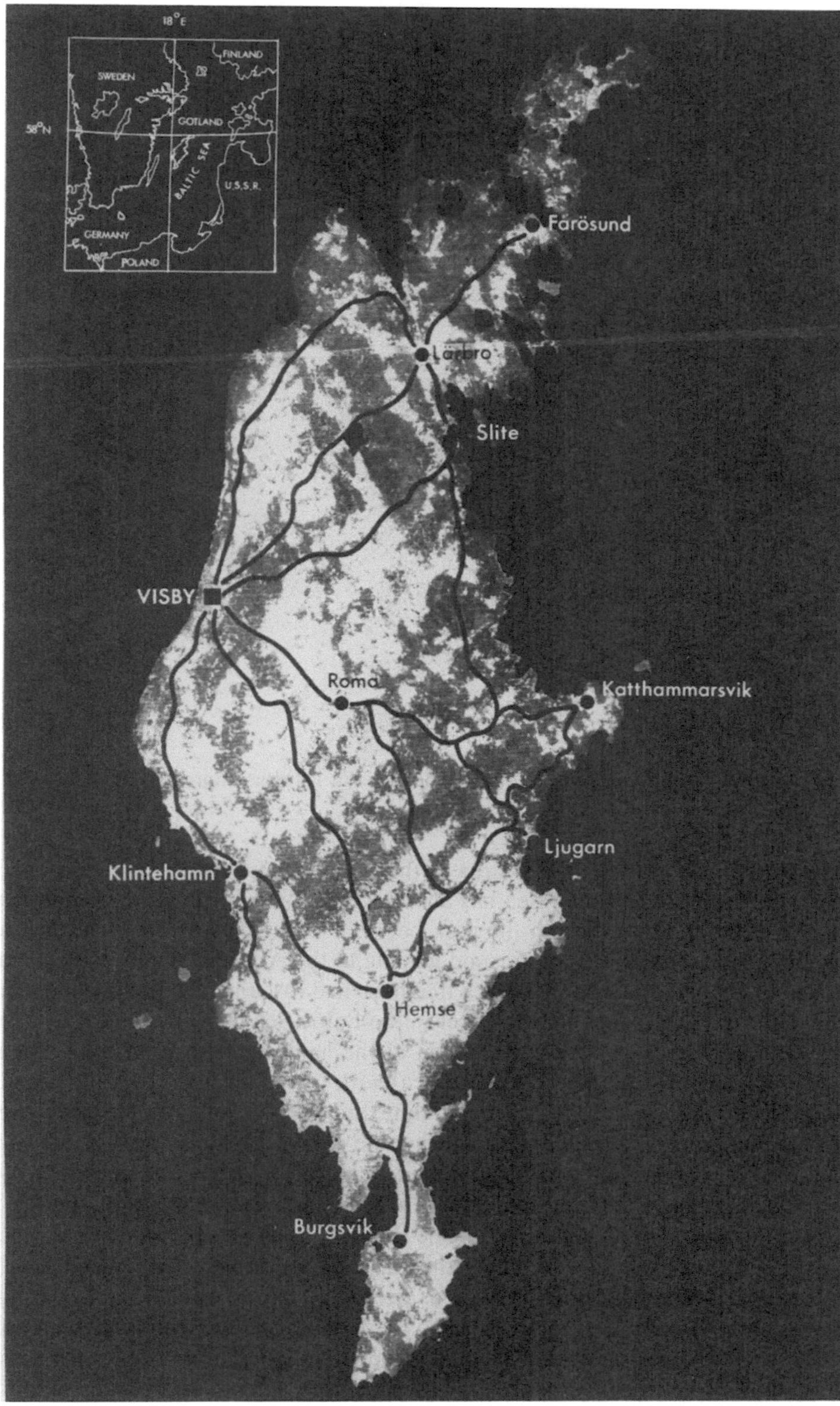
18°E
58°N
FINLAND
SWEDEN
GOTLAND
BALTIC SEA
U.S.S.R.
GERMANY
POLAND
Fårösund
Lärbro
Slite
VISBY
Roma
Katthammarsvik
Ljugarn
Klintehamn
Hemse
Burgsvik

Figure 2.2. Map of the Ancylus and Litorina shorelines.

in a continuous growth of land area of the island. However, since the last Ice Age, the level of the Baltic Sea has also fluctuated significantly. Variations in the level of land elevation in the area between Denmark and Sweden, now a channel called the Kattegat-Öresund leading to the Atlantic Ocean, have been linked to conditions in the Baltic Sea that have had resulting impacts on Gotland. Since the ice sheets melted, the water in the Baltic Sea has alternated between fresh and marine conditions. Elevation of the land resulted in the Baltic Sea becoming closed and fresh (Ancylus Sea) with the water level rising to about 50 m above the present level. Sharp delineations of the Ancylus shoreline can be seen on the island (Figure 2.2). During the Ancylus period, the climate was cold and dry, with tundra vegetation and scattered pine trees. Later (approximately 8000 years ago), the land between Denmark and Sweden sank, the Kattegat-Öresund channel was opened, and the Baltic Sea became saline (Litorina Sea), with the sea level also higher than at present; again, what was once the Litorina shoreline can now be seen as geologic features on Gotland. During this period, the climate rapidly became milder and more maritime, deciduous trees (such as oaks, elms, and lindens) invaded, and the tundra ecosystem with large flooded and swampy areas was succeeded by birch, aspen, and pine forests. The sum total of these comparatively recent geologic events led to a physical environment that was noticeable in terms of change. A Gotland chronicle from the Middle Ages, the Gutasaga, goes on to state: ". . . At the time the island was so enchanted that it sank by day and rose by night . . ." (Spencer 1974).

The first human immigration to Gotland took place during the late Stone Age about 7000 years ago. These Stone Age men were avid seal hunters and fishermen. The central position of Gotland in the Baltic Sea has always been used to advantage in trade, and prehistoric islanders were able to maintain trade with southern Sweden and Denmark. The Bronze Age, from about 1500–500 B.C.E., apparently was a flourishing period with a developed trade with Scandinavia as well as with Europeans to the south. Climatic amelioration and warmer weather enhanced development, with indications that trade occurred even with such distant areas as Greece. Impressive shipgraves up to 45 m long, as well as many different-sized cairns made by heaping up piles of stone, are remnants often found in close connection to the shoreline of that period. Harsher weather ushered in the Iron Age around 500 B.C.E., and it seems that a new people (possibly the Celts from central Europe) and culture came from farther south. A prosperous system existed from the beginning of the Christian era to about 500 A.D.; again, trade was indispensable to prosperity—with the Roman Empire, the Black Sea nations, and Byzantium. Impressive and organized cemeteries and graves, elaborate picture-stones (bildstenar), weapons, and more common implements are left over from this period.

From the earliest colonization up until 500 A.D., agriculture underwent various developments. The early dependence on hunting and fishing was supplemented by a slash-and-burn agriculture during the late Stone Age and early Bronze Age (2000–1000 B.C.E.) that used deciduous forests. Clearings were cultivated for 2 or 3 years and then left to recover for a 10-year period. Cattle, which were the basis of survival for the farm population, could be kept outdoors the entire year during this relatively mild period. During the climatic decline of the late Bronze Age and early Iron Age, the cultivation pattern intensified and became more stationary, with permanent settlements and stables required for the cold winters. Deciduous meadows became an important part of the farm from which hay, leaves, and branches could be harvested as winter fodder. Also, the swamps contributed fish and fodder to the farming household. The wetlands increased in area due to a period of damp climate and higher groundwater levels, which accelerated peat formation. From about 500 B.C.E to 200 A.D., each farm supported more than one family (Carlsson 1979); at the end of this period, however old territories were split into smaller units, pointing to a transition from a collective- to a single-family farming system.

In the middle of the fifth century, turbulent times became the norm. There is evidence of fire and violent destruction within the settlements—a time when the Holy Roman Empire was in a state of collapse. Gotland's inhabitants either fled or took refuge in strongpoints or forts such as Torsburgen in the southeastern part of the island, which was the largest fort constructed in Scandinavia. Some 200 years later, the area was influenced by Vikings, and a great increase in trade occurred as they established routes to Europe, the Middle East, the Near East, and Russia. Numerous findings of silver coins in Gotland's soils are an indications of this trade with eastern lands. The decline of the Viking Age eventually led to the conversion to Christianity in 1029 under the stimulus of the Norwegian King Olaf Haraldsson. For the next 300 years, Gotland flourished and experienced its "Golden Age," as trade expanded and the Germans and Gotlanders cooperated in building the capital, Visby, for the benefit of commercial enterprises. During this period, the walls around Visby were built, as well as numerous cloisters and churches in the capital and countryside.

Several developments arose that were to eventually undermine this period of prosperity. German merchants centered in Lubeck, employing tighter organizational methods, became competitors. Additionally, the success of Visby created a growing divergence in wealth between the city and the countryside; civil war ensued in 1288 and the countrymen were defeated. The Swedish King Magnus Ladulås established the terms of peace, and the organization of German merchants (Society of Germans Traveling to Gotland) moved to Lubeck, where it was to eventually become the Hanseatic League. Gotland, being a Swedish possession, became entangled in the disputes between Sweden and Denmark, and the Danish King Valdemar Atterdag conquered Gotland in 1361. The rural communities were seriously decimated, and Danish rule was to extend until 1645. This war led to a decline of rural society relative to urban Visby and, on the whole, a decline in living standards relative to the Golden Age. Many farms were abandoned during the fourteenth century (Ersson 1974), indicating a serious decrease in the farming population. At the end of the sixteenth century, as much as 13.5% of the farms were still deserted. Gotland became part of Sweden in 1645. Trade immediately increased with Stockholm, although a sometimes difficult relationship existed between the mainland and the island.

A new period of growth and development began after the introduction of new technologies and industries. Water wheels were built in small cataracts throughout Gotland. Later, wind technology was adopted and windmills rapidly grew in number (Figure 2.11). In 1697, there were 113 saw mills, 183 flour mills, and 48 windmills on Gotland (Moberg 1938). The export of raw materials (grain, limestone, firewood) slowly increased; as the manufacturing of goods expanded, new products emerged in the export trade. Lime burning, which had a long local tradition on the island, became industrialized when rich merchants settled on Gotland and invested money in large limeworks that produced for export. From the end of the seventeenth century and onwards, the annual export of lime amounted to approximately 20,000 m^3. This activity stimulated the overall level of employment and the economy of Gotland.

Valuable information about land use and agricultural production on Gotland about the year 1700 can be extracted from the map drawn by Swedish authorities as a basis for taxation. After Gotland was recaptured by the Swedes in 1645, the immediate concern of the Swedish government was to try to improve agricultural production on the island. The yields were found to be much lower than in other areas of Sweden. Experts were sent to Gotland to investigate the local conditions. They reported that natural factors, such as thin soils, wind

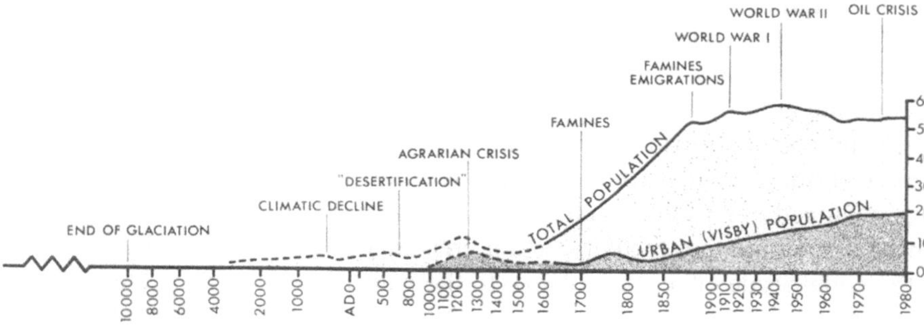

Figure 2.3. Long-term view of development on Gotland from prehistory to the present. Solid curves were constructed from actual population statistics. Dotted curves are reasonable estimates based on historic documents, archaeologic evidence, and climate change.

erosion, dry summers, and spring and autumn flooding, were the main reasons for the low yields. The total area of arable land was 195 km^2 in 1700 (Lindquist personal communication) but one-third to one-half of that area seems to have lain fallow since a 2–3-year crop rotation system predominated (Moberg 1938).

The nineteenth century was characterized by large-scale reclamations of land, as well as by the reorganization of property holdings. A drainage program was begun in 1820, which along with an extensive clearing of forests increased the amount of arable land by approximately 470 km^2 during the nineteenth century. In 1805, the harvest of grain was about 80,000 barrels (corresponding to about 10,000 tons), 10% of which was exported. One hundred years later, the harvest had increased to approximately 50,000 tons, of which 15,900 tons (or 32%) was exported (Olofsson 1945).

The increasing exploitation of natural resources during the nineteenth century was accompanied by rapid growth of the human population (Figure 2.3). The excess of births over deaths was high during the entire century; but compared with the rest of Sweden, the nativity was low on Gotland (Olofsson 1945). Between 1800 and 1880, the population increased from 31,000 to 55,000. Several years of failing crops then caused a boom in emigration. In the 1880s about 6000 persons (10%) emigrated to the Swedish mainland or to North America. At the same time, people moved from rural areas to Visby and other small towns. In 1900, the urban population was 8400 compared to 4600 in 1850. Due to increasing industrialization and trade, the level of income and purchasing power had markedly improved until the beginning of the twentieth century.

The Industrial Revolution has successfully transformed the economy into a modern and diverse one. The establishment of railways enhanced the importance of Gotland's agriculture to Stockholm, food processing industries emerged, and the steam-powered saw led to a booming timber industry supplying timber for structural support in the coal mines of Great Britain and the Ruhr. Fishing expanded; although it now constitutes a small proportion of the total economic value, the catch of herring, salmon, flounder, and cod represents a significant value in terms of protein. The natural and managed ecosystem-based activities of agriculture, forestry, and fisheries contribute substantially to the present-day economic activity on the island.

Since the end of World War II, several developments have been initiated that have fundamentally affected present conditions on the island. A great change occurred due to the extensive mechanization of agriculture. This led to enhanced agricultural productivity, and it created a surplus of labor that migrated from the countryside either to Visby or to the mainland of Sweden. The absence of sufficient employment opportunities on the island resulted in a net emigration, and the island population dropped from about 59,000 in 1945 to a present population of about 54,500. Other activities were developed that counteracted this lack of available employment. Tourist activity grew substantially and was stimulated by the first car ferry in 1955, two new car ferries in 1964, and a very large, ultramodern ferry introduced in 1981 that facilitated access to the island. The locating of a branch of the L.M. Ericsson telephone company and a State Lottery headquarters in Visby during the 1960s, the expansion of other industrial activities (e.g., cement production capacity is now double what it was in 1972), and the recent expansion of fish-processing plants at Visby all have contributed to employment opportunities. This has reduced the incentive to migrate to the mainland, with a reasonably stable population level as a result. In addition, military activities generate income and jobs for the economy.

Many of these activities were facilitated by electrical energy available from an underwater cable, which was connected to the mainland in 1954 and whose capacity was extensively expanded in 1983. Thus, the emergence of post-World War II developments, which led to a declining population, were balanced in many ways through deliberate regional and national policies designed to stem the flow of population; a reasonably stable and, in fact, slightly increasing population seems to have been established. This long-term view of historic development is presented in Figure 2.3, which indicates major natural and human events and associated population levels from the earliest settlements. Table 2.1 indicates measures of development since 1700. Note the extent to which the rural-to-urban proportion has changed and the extensive increase in agricultural land, inputs, and production. Although the total number of cattle has increased, the density of cattle was much higher in the eighteenth and nineteenth centuries compared to the present. The agricultural information for Table 2.1 was obtained from Olofsson (1945). Our reconstruction of the long-term development of population is subject to error because of the lack of hard data prior to the sixteenth century. We know that Visby had about 6000–8000 inhabitants in the twelfth and thirteenth centuries and that the population declined after the agrarian crisis until the sixteenth century (Gotlands Kommun 1974). Archaeologic evidence also indicates a regression of society on Gotland, and all over Europe, beginning in about 600 or 700 A.D. (Carlsson 1979). Furthermore, it seems that around 200 A.D., the shift from extensive space demanding exploitation to smaller farms and a more intensive use of agricultural land led to an increase in population. The family or extended family as a social unit became more important

Table 2.1. Some Measures for Gotland's Agricultural and Urban Development[a]

	1700	1800	1850	1900	1950	1960	1970	1980
Population	18,000	31,000	44,600	53,000	59,000	54,000	54,000	55,000
Rural/urban population	6.2	7.5	9.7	5.3	1.6	1.0	0.9	0.9
Arable land (km^2)	195	150	244	650	830	810	815	830
Arable land (ha/capita)	0.6	0.5	0.5	1.2	1.4	1.5	1.5	1.5
Fertilizers (kg/ha)	—	—	—	46	370	580	650	600
Yield (grain) (tons/ha)	?	1	1	1.2	1.9	2.4	3.4	2.0
Cattle	15,000	20,000	19,000	25,000	43,000	50,000	50,000	57,000
Cattle/km^2 arable land	150	133	78	38	52	61	61	67

[a] Jansson (1985).

and private ownership became predominant. These developments were interrupted by the violent period of the seventh or eighth century A.D. and by an accompanying population decline.

General Description of the Existing System

Natural Systems

Gotland has a remarkable diversity of features because of its limestone based geology, natural ecosystems, and long history of habitation, which has left numerous impacts on the landscape in the form of structures and land use patterns. The limestone-based geology creates a landscape that is very different from the mainland, which is an important attraction for tourists. Transportation is now facilitated by a network of roads. Since the island is small in area—only about 150 km long and 50 km wide at its greatest width—it is a short trip to traverse it. However, an exploration of the many diverse villages, churches, historic sites, and interesting fauna and flora can take considerable time. Many of the beaches are juxtaposed next to sharply rising limestone cliffs on the landward side; this is especially characteristic of the northwest coast as one travels north from the capital, Visby. These cliffs contain many caves, some of which are used as a tourist attraction at Lummelunda. Imagine being on a limestone cliff at the shore, perhaps 50 m high, with thin soils and a gaunt and sparse heath landscape, the gray Swedish light of autumn or winter, and winds, clouds, and waves indicating an approaching storm; one might have some feeling about the impression of some of the Gotland landscapes. These coastal landscapes become even more enchanting in the areas of marine stacks (raukar in Swedish) left over from eroded relics seabeds. On the northwest coast of Fårö, raukar are abundant and are composed of remnant fossil coral reefs that have withstood the weathering, taking on odd and fantastic shapes for one's imagination to transform. There are, of course, many other types of shorelines along the coast, including extensive sandy beaches.

At a latitude of about 57°N and longitude 19°E, Gotland is near the extreme northern boundary of the inhabited world. However, its weather, as with all of Scandinavia, is moderated by the Gulf Stream; furthermore, the presence of the Baltic Sea smooths out much of the seasonal extremes. The mean temperature at Visby of 7°C is about 1°C higher than at a similar latitude on the Swedish mainland (e.g., Jönköping). The growing season is quite longer, and many plants are found that do not grow in other areas at the same latitude. Precipitation is rather low, averaging about 525 mm/yr with little rain in the summer. The mean highest monthly temperature is about 27°C in July, whereas the mean lowest temperature is about −10°C in February. The coastal water surface temperatures do not get much higher than 17–18°C in late summer. The annual number of hours of sunshine is the greatest for all of Sweden. Summer and early autumn are sunny for this latitude, attracting many tourists from the Swedish mainland.

The island as a whole is a low-relief, undulating limestone plateau that is occasionally marked by some high points (highest point, 83 m), plateaus, and jagged cliffs along some of the coasts. Gotland is subject to winds that traverse the wide areas of the western Baltic Sea. The average annual wind speed is approximately 6.2 m sec^{-1}, with highest velocities during the winter. The high winds bring waves with great force on the west and northwest coasts. Wave energy on this side of the island tends to be expended quickly and nearer to shore in relation to the east coast, since the depth of the sea bed increases much more rapidly (on the northwestern side, the 30 m depth isopleth is about 0.8 km from shore versus 8 km off the eastern shoreline).

Away from the coast and in the interior is a diversity of natural landscapes whose productivity is highly dependent on the nearness to available supplies of water and soil depth. Before the extensive drainage, which began in the nineteenth century, there were some 300 km^2 of wetlands in comparison to the approximately 40 km^2 that exist now. These drained areas have provided rich peat soils, and they are the sites of many of the most productive farms. Agricultural production is one of the main economic uses of the land (about 810 km^2 is cropland and grazing land) as is forestry. Travelling across the island will reveal these two primary land-use types organized in a patchwork fashion; the pattern depends on soil suitability for agricultural conversion (see Figures 2.4 and 2.5). The drainage of wetlands has led to a lowering of the water table, with water becoming a limiting factor in many areas, especially for the natural systems. It is always striking to go into forested areas where the water table is high and to note the lushness and productivity of the vegetation in comparison to a similar stand located in a drier area.

At present, the trees that dominate are pine (*Pinus silvestris*) and spruce (*Picea abies*). Deciduous woods are relatively rare except in the south, since the long agricultural history of the island has limited broad-leaf trees in small groves in fields and in areas adjacent to farmhouses, churches, roadsides, and forest edges. The trees and shrub flora are dominated by esher (*Fraxinus excelsior*), hazel (*Corylus avellana*), oak (*Quercus robur*), elm (*Ulmus glabra* and *U. carpinifolia*), and birch (*Betula pubescens* and *B. verrucosa*). The rich herbaceous layer is noted for many species of orchids flowering in spring and early summer.

A cultural variation of the deciduous system is the wooded pasture "änge." These "ängen" played an important role in fodder production prior to the introduction of cultivated clover pastures. At present, only a small percentage of the original änge area remains to be maintained. The typical änge is a mosaic of open glades, dense groves, and shrubs maintained by human harvesting and animal grazing. In winter, the woods are thinned and cleared; in spring, litter is raked and burned. After haymaking in the middle of summer, cattle are allowed to graze for a couple of weeks. At the end of the growing season, trees and bushes are used for harvesting with their leaves and branches and are stored as winter fodder for animals. This human activity prevents a natural succession of bushes and coniferous trees, leads to a high diversity of ground vegetation, birdlife, and other associated animals and—at the same time—preserves the old cultural traditions of work and feasts.

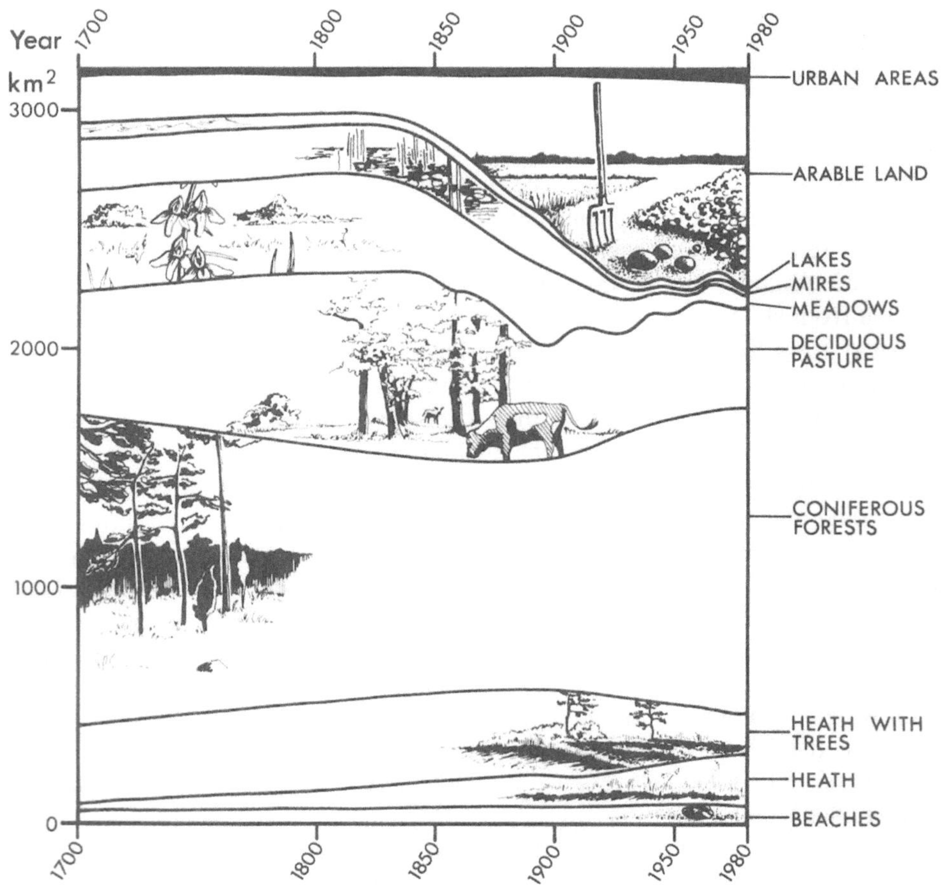

Figure 2.4. Estimates of the change in land use from 1700–1980 (see Appendix I) (Jansson 1985).

Heathland, both with and without trees, is another distinctive part of the landscape. When the ice and sea withdrew from the island after the last glaciation, vast flat areas of limestone were left with little or no sediment on top. There still are large areas of almost bare bedrock, with a sparse vegetative covering of lichens, mosses, and *Sedum* species. *Juniperus* shrubs, grasses, and herbs with scattered pine and sorb trees can be found where weathered soil has accumulated. On well drained deeper soils (20–30 cm) on glacial till or gravel, open pine forests are characteristic. Although heathland productivity is low, its contribution to the total system cannot be ignored. Heathlands have been extensively used for firewood (especially during the period of intensive lime burning in the nineteenth century), are used for grazing by sheep, and support wild populations of rabbits, hares, and foxes. The low tree density and high solar intensities at ground level result in a high diversity of flora in many of these areas. Proposed tall cement towers supporting wind-powered electrical generators may, in the future, add a surrealistic quality to some of these landscapes.

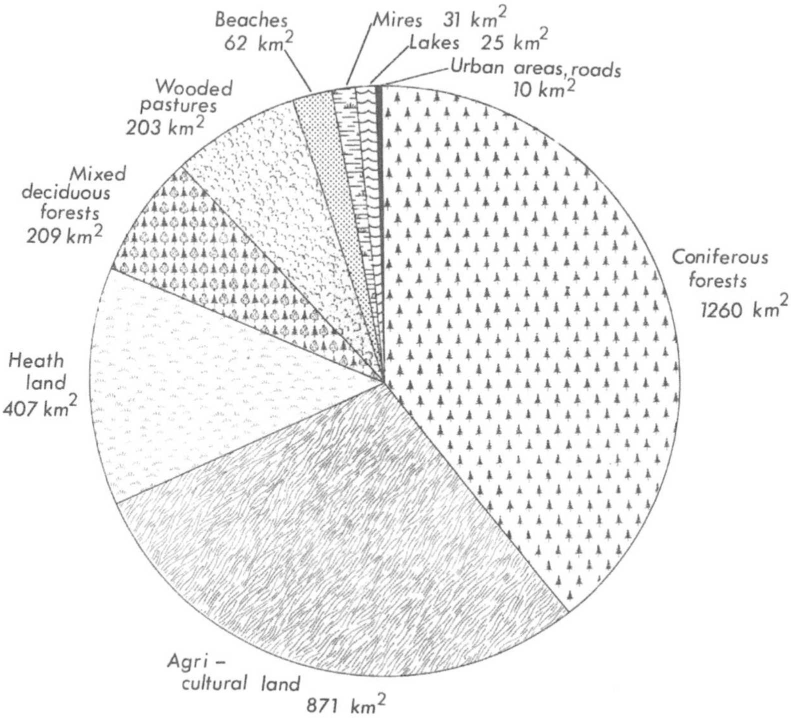

Figure 2.5. Present-day distribution of land use (see Appendix I).

Mires and wetlands were historically much more prevalent, and they constituted 10% of the total land area as late as 1820. Peat layers of 0.5–6 m have developed in shallow basins of bedrock, where flooding inhibits complete decomposition of the organic matter. Vegetation in the mires is dominated by *Salix* shrubs, *Cladium mariscus*, and *Carex* species. The mires played a diverse role in ancient local farming, serving in winter and spring as freshwater lakes for fishing and in summer for hay harvesting and livestock grazing. *Cladium* vegetation was harvested for use as long-lasting roofing material. Moreover, the mires played an important hydrologic function, providing a freshwater magazine that served as a vital buffer to shortages of water during the growing season.

A final category of terrestrial ecosystem to consider is the seashore. Four main types are found around the island: rocky coast, gravel beach, sandy beach, and beach meadows. Many of the features of the beach systems were created during the different transgression periods of the Baltic Sea, by the action of winds, and by the work of wave erosion; the characteristic klints and raukar comprise one distinctive feature of the beaches that are sculpted by winds and waves. Photosynthetic production on most of the beach systems is small, except for the more protected beach meadows where great masses of nutrient-rich algae accumulate. Here, production is high enough to feed cattle and sheep. Beach meadows also support rabbits and birds, and they provide places for resting and moulting for tens of thousands of migrating geese each year (Stolt 1971, Nilsson 1975). The vegetation is eaten by the animals who, in return, excrete nutrients into these areas; this constitutes a "feedback" by the grazers to the system that supports them. The rapid increase in tourism has stimulated the building of summer houses, hotels, and camping lots in areas near the shore. Although the immediate beach areas have been protected from private housing by a 1948 law, there is a high density of small summer homes within 500 m of the seashore. The beach systems continue to be an important resource for tourist activity and Gotland's economy.

The freshwater systems are much diminished in importance since the advent of large-scale drainage. Many of the remaining small creeks dry up completely in summer. The most important lake is Tinstäde Träsk, which supplies water to Visby. Dense *Phragmites* and *Cladium* stands along the lake shores support large bird populations. On the other hand, the surrounding Baltic Sea is important to the island in many ways. It ameliorates the climate, supports fisheries, assimilates pollution, may provide sand for industry, and furnishes wave energy for potential future energy conversion. Down to a depth of 7–10 m, the bottom vegetation consists of *Fucus vesiculosus* (bladder wrack) on rocky bottoms and *Zostera marina* (eelgrass) on sandy bottoms, with red alga belts at lower depths. Bottom fauna is dominated by plankton-filtering blue mussels, small crustaceans feeding on dead organic matter, and bacteria. The gross primary production of the coastal systems provides energy to support populations of flounder, herring, cod, salmon, seals, and sea birds. Because of available ice-free waters, numerous wintering waterfowl gather around Gotland (Nilsson 1975); estimates are 30,000 long-tailed ducks in midwinter and total summer populations of waterfowl amounting to 50,000–60,000, with extensive feeding both in nearshore areas by wading birds and in offshore areas by gulls, terns, guillemots, and ducks (Stolt 1971). Although birds compete with fisheries, exploiting almost as much fish every year, they also contribute to the recycling of nutrients and supply terrestrial systems with additions of phosphorus. Human activity, such as land runoff, sewage outlets, oil spills, toxic substances, intense fishing, boat traffic, and dredging, have various impacts on the coastal ecosystem; still, the waters around Gotland remain in fairly good condition compared to many other parts of the Baltic Sea.

Human Systems

The beauty and diversity of the natural landscape has been affected in many ways by the existing and former presence of human activities. If one takes a ferry from the Swedish mainland to Visby, a welcome and enchanting sight emerges as one approaches this main town. Because of the long period of habitation, there is a rich variety of architectural forms in Visby ranging from the old to the new. In addition, the beauty of the town is enhanced by the fact that it is built on a series of narrow limestone plateaus, mainly on three levels. This vertical dimension adds to the view, either looking out from the city or towards the city from the sea. The central part of the city is surrounded by a wall built around 1100 A.D. that is still in excellent condition along most of its structure; it was used as a defense for the city, especially against the countryside. Within the wall are well-preserved buildings dating back to the Middle Ages with many German architectural influences. The cobblestone streets are narrow, buildings are low and compact, and walking is always interesting as streets ascend, descend, and run at different angles to one another. Many of the houses have neatly kept gardens colored with flowers in the spring and summer; museums and shops provide glimpses into past as well as present activities. A visit to Visby in the winter will present a quiet and relatively empty and quaint city, whereas the summer throngs of tourists completely change the atmosphere with noise from private cars, bicyclists, souvenir stands, open-air restaurants, cafes, and tourist guides. The population of Visby has expanded over the past 20 years, from 15,000 to about 20,000 inhabitants. The new growth has occurred mostly outside the walls where the architecture is modern. Rows of apartment houses or private homes, supermarkets, and shopping centers remind one of the mainland and a different world than that suggested by the old city within the walls (see Figure 2.6).

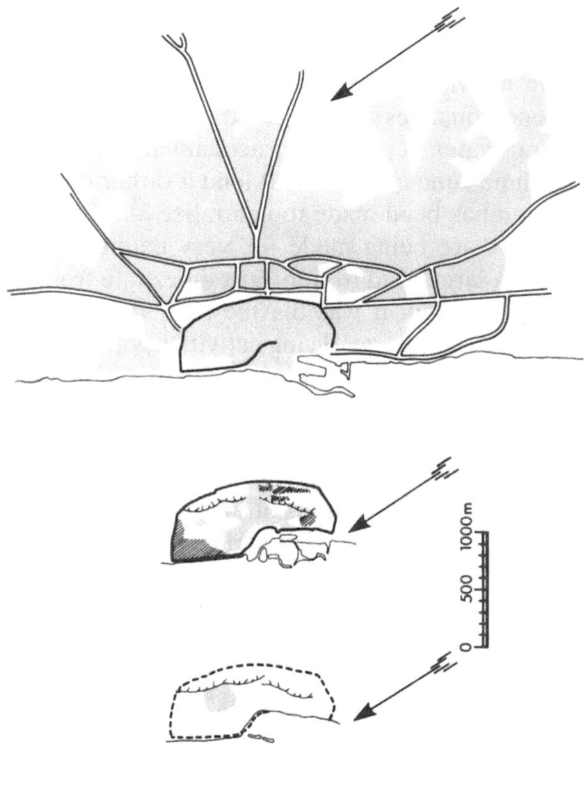

Figure 2.6. Map of old and new areas of Visby. The bottom figure indicates the small Stone Age settlement, the middle figure the Middle Ages, and the top the present. The wall built in the Middle Ages is indicated by the heavy line. The bottom two figures are from *Gotlands Kommun* (1974). The top figure is constructed from a 1979 road map.

As one leaves Visby and the extensive tourist camps along the coast, the countryside—with its mixture of natural landscapes, farms, and forests—is rapidly reached. There are many towns scattered across the island. Most of these serve as agricultural focal points for the surrounding countryside, although each has its unique qualities and character. Some of the towns are influenced and dominated by a prevailing industry, such as the cement factory in Slite, the sugar factory in Roma, or the military unit in Fårösund. Slite bears the familiar marks of a small industrial town; a large cement company (Cementa) and the power company (Gotland's kraftverk) lie very close to the harbor. Picturesque remnants from former times include almost 100 churches scattered across the countryside, which date back to the Middle Ages and are still in good condition, as well as the abandoned structures of old windmills and a few water mills that were used extensively in former times. They add a remarkable diversity and uniqueness to many of the areas, as do other sites such as old farm buildings and grave sites that include mounds and the spectacular ship-graves. A grave site of particular enchantment because of the surrounding, rather moist forest is located at Trullhalsar on the east coast. Much of the northern parts of Gotland and up towards Fårö are used by the military for exercises and maneuvers. Also, small oil pumps have recently been set up in this area to try to obtain a flow of oil from recent discoveries of small pools 200 m below the ground.

One of the main activities that has existed for a long period is limestone quarrying. Abandoned quarries are scattered across the land, with some vegetation invading and water beginning to accumulate at the bottom. Close to Slite are extensive limestone quarries that lend a rather desolate feeling to the area. Since a decision has been made to centralize all of Swedish cement production at Slite, plans are being made for very extensive new quarries just outside of the town. A large road to transport limestone from the quarry to the factory has been constructed; it was blasted right through limestone cliffs in some sections. These extensive quarrying activities will eliminate a biologically diverse terrestrial community on the site, and they may also affect the hydrology of the area. Other large quarries exist at Storugns in the northwestern part of Gotland; at this point, rock is shipped out for export. These activities—together with the drainage of wetlands, the large-scale clearing of timber in the nineteenth century, and the extensive fertilization of agricultural lands leading to water quality impacts—exert the major detrimental environmental impacts on the island. On the whole, however, because of its relatively low population density, Gotland continues to have a diverse rural landscape.

Macroregional Analysis

In order to gain some overall quantitative perspective on the total system and the magnitudes of activity, gross regional data pertaining to population, energy, economic activity, and critical resources are presented in this section. As discussed in Chapter 1, macrolevel data can be used to make comparisons to other regions or to identify trends and changes over periods of time with which prognostications can sometimes be made easily. A comprehensive study of a region should consider detailed flows of material resources throughout the system, but this can become overly cumbersome if some assessment is not made of material flows in terms of importance; to this end, those resources are discussed that are deemed most critical to Gotland at least for the next 20 years or so.

Population

The total population of Gotland is relatively small; it was about 54,800 in 1978, with an average population density of 17.5 persons per km^2 (45.2 persons per square mile). Approximately 43,000 inhabitants reside in the urban areas, of which the towns of Visby, Fårösund, Hemse, Klintehamn, Roma, and Slite are the most important regarding numbers. Visby is the capital of the municipality of Gotland, and it constitutes the major urban area. The rural population amounts to about 12,000 people. The distribution of population has changed considerably since World War II, with a decline in the countryside and an increase in the urban areas. The population as a whole declined during this period because of emigration (Figures 2.7–2.10) due to surplus labor created in the aftermath of agricultural mechanization; there were not enough alternative jobs to absorb this labor force. It can be seen that emigration from the countryside exceeded immigration during this period, with both decreasing towards the 1960s while the population of Visby increased sharply since the late 1950s. The population age distribution for the entire island as well as for Visby has also changed from 1940 to the present; in the past, age structure was characterized by a pyramid with a larger base and more members in the younger age brackets (Figure 2.8). The changes that have occurred since then have led to a more even distribution of population by age. Younger people have migrated and older people have tended to remain on the farms.

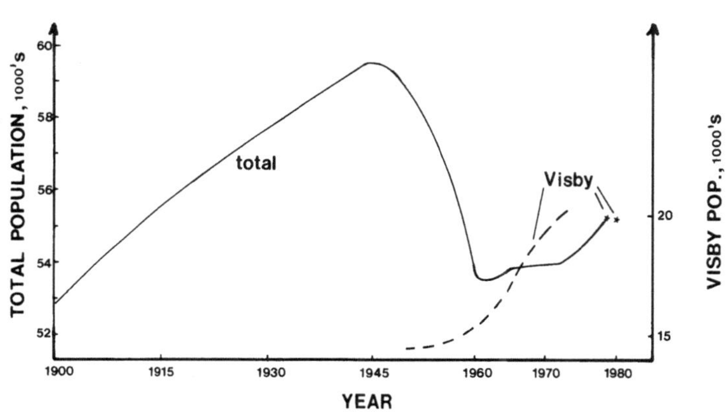

Figure 2.7. Total population of Gotland since 1900 and the rapid rise in the population of Visby in the post-World War II period. The population data is available from official statistics of Sweden.

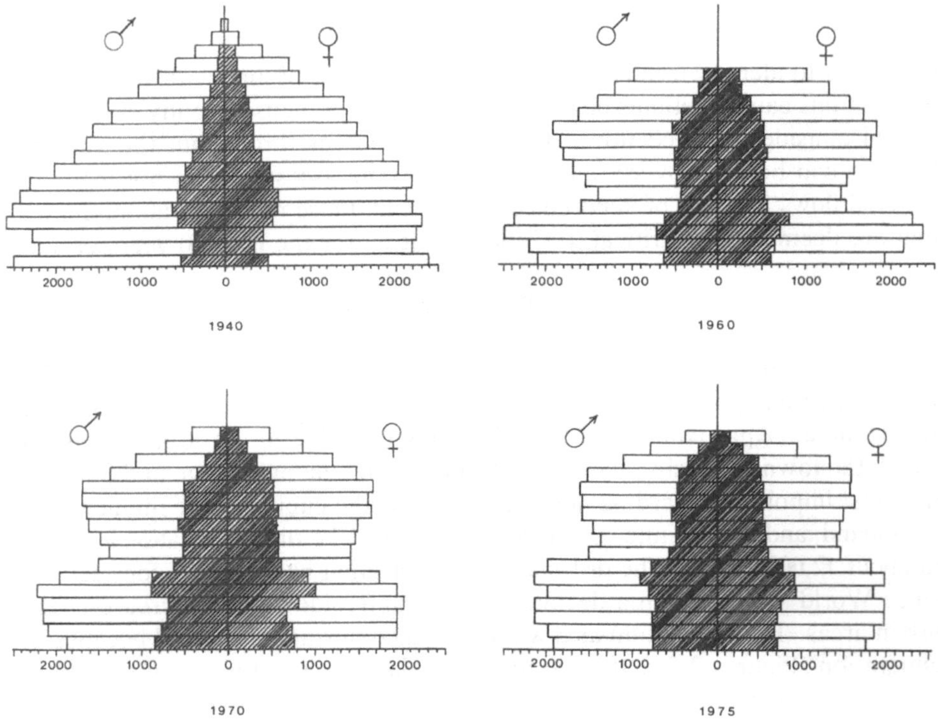

Figure 2.8. Age distribution of Gotland's population by 5-year age brackets. The darkened bars designate the Visby population.

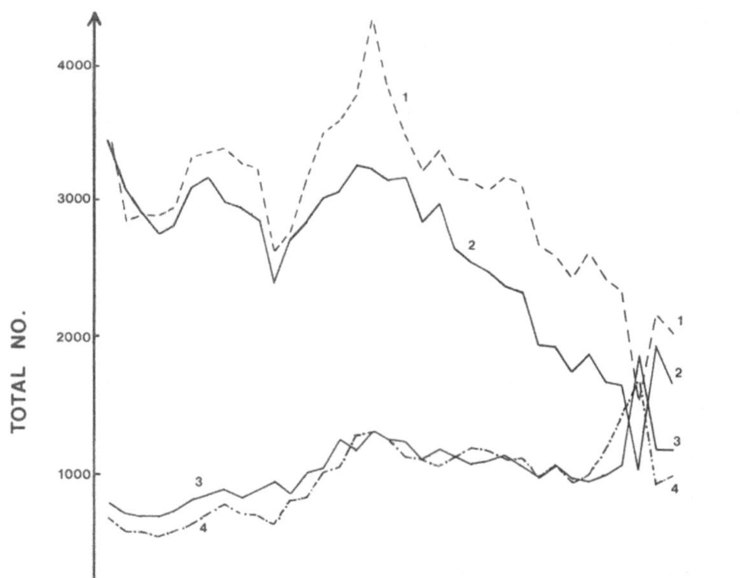

Figure 2.9. Some trends in emigration and immigration for the cities and countryside of Gotland. (1) Emigration from countryside. (2) Immigration to countryside. (3) Immigration to cities. (4) Emigration from cities.

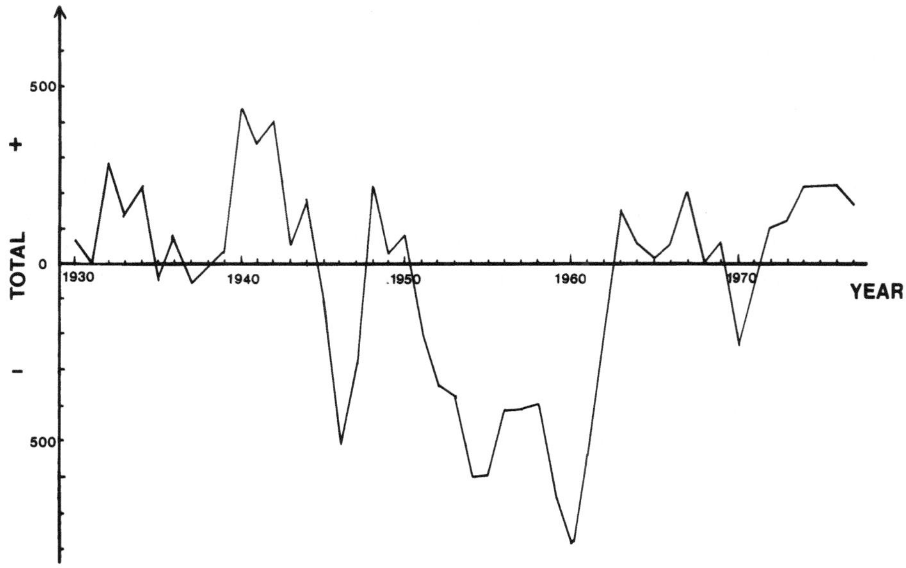

Figure 2.10. Increases and decreases in Gotland's population.

It is interesting to note the alternating periods of population increase and decrease, much of which were due to migration (Figure 2.10). Increases seem to have occurred during periods of stress in the country as a whole, such as the Great Depression in the 1930s and World War II. The recent immigration during the 1970s may be partly due to the slower growth of the Swedish economy as well as to the "green-wave" movement away from metropolitan areas. The sharp decline during the 1950's may have continued if there were not deliberate national efforts to support declining regions. The institution of incentives to encourage industries to locate on Gotland and of subsidies to transportation to and from the island, and the expansion of government offices, all created local opportunities for employment and arrested the decline. Since the 1960s, tourism has also helped to create job opportunities. The population has remained remarkably constant since about 1963, even though the past few years have witnessed a steady immigration.

Population projections obviously are important for any regional study attempting to assess energy, resource, and environmental ramifications of human activity. Certainly, one could build a population cohort model with migration effects built in; however, it is still difficult to determine the extent to which external events influence migration. We have not engaged in building a population model because of the relatively steady-state situation that has existed in population levels over the past 20 years. This has facilitated our analysis considerably in eliminating the uncertainties associated with a rapidly changing population—either growing or declining.

This population stability is not manifest, however, if the total number of tourists visiting Gotland is considered. Since 1950, the number of tourists has steadily risen and is now 300,000 per year, which is a large number for Gotland since most of these visits are made in June, July, and August. This large influx of tourists brings economic benefits, but it also generates costs because of the need to provide infrastructure, maintain order, and provide services and resources. What is it about Gotland that attracts tourists? The diversity of human and natural surroundings, the difference of the landscape in comparison to the mainland of Sweden, and the availability of space along the coastlines all contribute to the creation of an "image" that attracts many people. Equally important are the means of transportation, which are reasonably priced, to the island; the increased availability of large ferries shuttling back and forth from the mainland to Visby has clearly affected the number of tourist visits.

Energy Overview

The region of Gotland depends on the renewable energy flows associated with solar energy as well as on fuels and electricity. There are several couplings between the renewable solar-based energy flows and the fuels and electricity (purchased energies) involved in the human economy; agriculture, forestry, and fisheries are the most obvious examples. In this section, an overview of the energy flows on the island is presented, as well as time series data to identify trends and changes that have taken place. The natural energy flows that have prevailed as long as Gotland has existed as a terrestrial system are those of sunlight, rain, winds, waves, gravitational potential, chemical free energy of mixing, and geologic processes. Although somewhat different in the past, the magnitudes of these flows presently are reasonably constant over the human time frame; however, variations occur from year to year. The early inhabitants of Gotland were completely dependent on these natural energy flows, which supported their agriculture and fisheries. The sun provided warmth and fresh water through the hydrologic cycle, while the winds were an energy source for the ever important transportation to and from the island. Later on, new technologies in the form of windmills, wind saws and wind pumps harnessed wind energy more directly for performing work; there were about 500 windmills dispersed over the countryside in the beginning of the twentieth century (Figure 2.11). These renewable energy flows are still important to the fundamental energy basis of the island, since agriculture, forestry, fisheries, and tourism are vital economic activities of the human system; however, they have been augmented greatly by fuels and electricity.

One of the first undertakings of our project was to calculate the magnitudes of the natural energy flows (Table 2.2). They represent significant quantities of energy, at least in terms of actual energy flux, but their ability to do work depends on their quality and the available technologies. The largest flux, 20,400 PJ yr^{-1}, is solar energy and is diffuse. Much of this energy is either reflected or dispersed in the environment as heat. Some of what remains is used by the dispersed biochemical pathways of the photosynthetic systems on land and offshore. Of this flux, only 100 PJ yr^{-1} (0.5%) is channeled into gross photosynthetic production of agriculture, forests, deciduous glades, lakes, mires, heathlands, beach systems, and Baltic coastal systems. This stored chemical

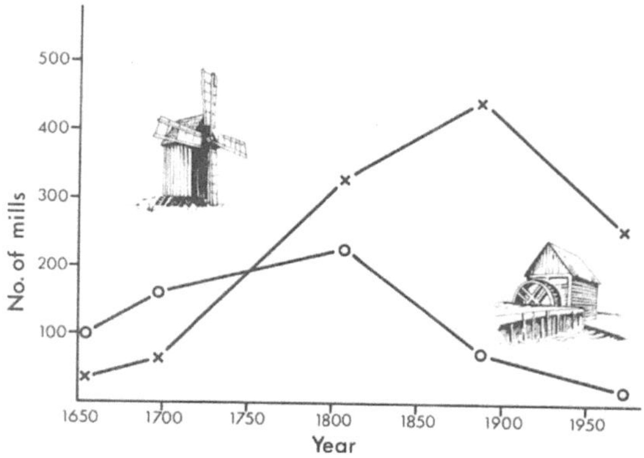

Figure 2.11. Number of windmills (×) and water mills (o) from 1650 to recent times.

energy produces growth, maintains diversity of flora and fauna, drives cycles of material flow such as water or nutrients, provides services for humans, and provides goods in the form of harvestible food, fiber, and materials. The solar flux during summer is especially important, since it is responsible for attracting tourists.

Gotland is subject to fairly strong winds, especially during the winter season, with average wind speeds of about 6 m sec^{-1}. The total wind energy of 152 PJ yr^{-1} within 100 m of the ground consists of horizontal movements of air that sweep across the island, as well as vertical down-welling of air from upper strata through turbulent diffusion. Wind energy is also converted into wave energy, which is dissipated along the coast through the medium of the Baltic Sea. The total wave energy of 56 PJ yr^{-1} contributes to beach maintenance, nutrient transfers, and turbulent mixing in the coastal systems. The chemical potential energy of fresh rainwater (6 PJ yr^{-1}) represents the minimum energy required to convert brackish Baltic water (salinity = 6 ppt) to fresh water; this

Table 2.2. Estimates of Natural Energy Flows on Gotland, Sweden[a]

Category	Area (sq km)	Total Energy Flow (TJ yr^{-1})
Solar energy	3140	20.4 million
Gross production of terrestrial systems	2170	57,000
Gross production of Baltic coastal systems	2450	17,600
Agriculture	870	27,700
Wind energy (0–100 m)	—	152,000 (maximum)
Beach wave energy	—	56,000
Free energy of salt/fresh water gradient	—	6000
Potential head of water	—	200
Geologic uplift	—	

[a] Calculations can be found in Jansson and Zucchetto 1978a. See also Appendix II.

work is accomplished by the evaporation of seawater through the expenditure of solar energy in other regions. Rainfall on elevated areas has the potential for doing work because of its gravitational potential energy, with respect to sea level. As this water flows downhill it creates rivers, distributes nutrients to both land and sea, and can be used by humans if hydroelectric facilities or other mechanical devices are employed. In many regions, hydroelectric potential may be extremely large; but on Gotland, this potential energy is quite small (0.2 PJ yr^{-1}) due to its flat topography, low rainfall, and absence of large rivers. One final category of natural energy release involves geologic work, as Gotland rises at a rate of about 0.25 cm yr^{-1}. This postglacial rebound is due to the expansion of the underlying strata, which were compressed by the weight of ice sheets during the most recent Ice Age. This land rise is creating new land at a slow rate, and it represents a large flux of energy whose magnitude is difficult to calculate. The extent to which some of these natural energy flows can be used by new technologies is addressed later in this chapter and in Chapter 3.

With the presence of a human society, and its ability to import and use fossil fuels as well as to produce electricity, the energy regime of the region has been significantly changed over time (Figure 2.12). Following wind and wood, coal was used as a more abundant and concentrated form of energy from the late

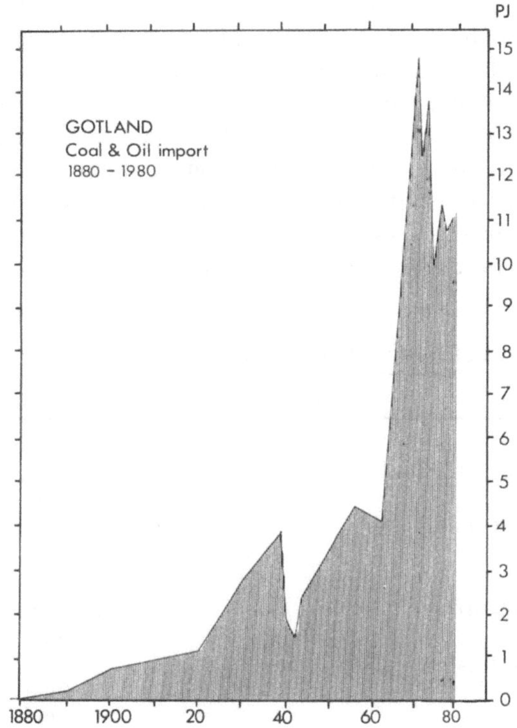

Figure 2.12. Historic view of total annual energy consumption on Gotland (data obtained from Olofsson 1945 and A.B. Svenska Shell in Visby).

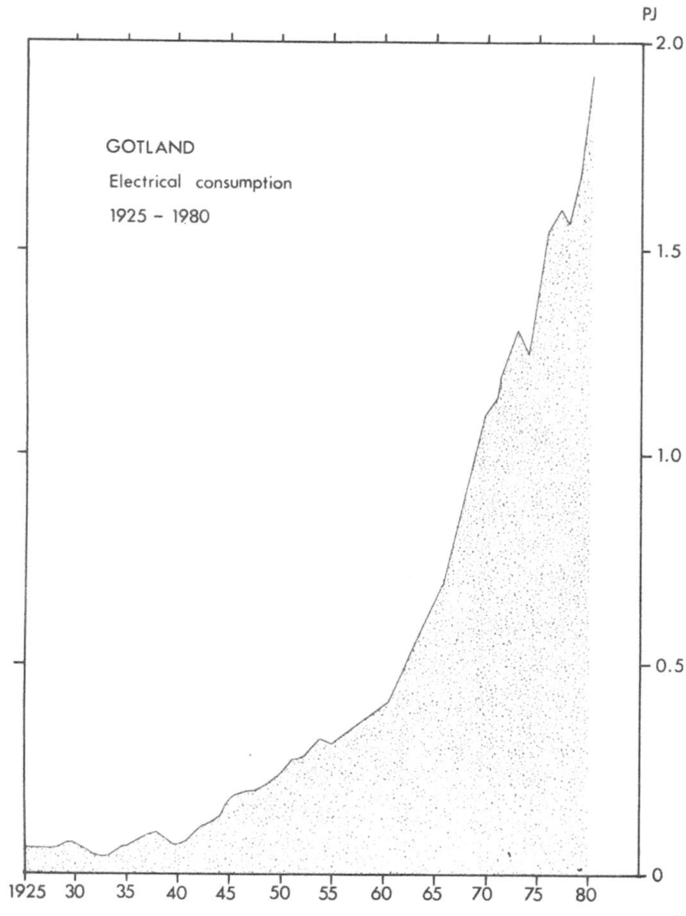

Figure 2.13. Historic view of electrical energy consumption on Gotland from 1925–1980 (data obtained from Visby Elverk and Gotlands kraftverk).

nineteenth century until its relatively recent replacement by oil. The growth of electrical energy consumption has also been pronounced since 1925, with rapid rates of growth since the early 1960s (see Figure 2.13). Electrical generating capacity is about 111 MW, while a cable from the mainland built in 1954 serves as an additional source of high-quality electricity; it has now been expanded from 30 MW to 130 MW. This cable can be used for both import and export of electricity. During periods of low hydroelectric capacity or a high demand for electricity on the mainland, excess generating capacity on Gotland can be used for supplying the mainland. Gotland is almost totally dependent on imports of petroleum fuels, electricity, and (in the near future) coal. The indigenous use of wood fuel accounts, at the moment, for a small amount of the total energy consumption. Surprisingly, oil was found in the late 1970s and small wells are continuously pumping, mostly in the northern part of the island. Estimates are that crude oil production is about 50 m^3 day^{-1} (18,000 m^3 yr^{-1}) in

Table 2.3. Summary of Major Components of Imported Energy, 1970–1978

	Gasoline		Diesel		Oil (No. 1, 3, 4, 5)		Other Fuels		Electricity		Coal	
Year	m^3	TJ	m^3	TJ	m^3	TJ	m^3	TJ	GWh	TJ	Tons	TJ
1970	31,465	988	20,600	733	313,804	11,093	31,156[a]	1090	40	142	—	—
1971	31,569	991	19,150	682	288,732	10,492	3707	131	127	456	—	—
1972	32,027	1006	20,200	720	313,844	11,881	6333	221	154	555	—	—
1973	32,675	1026	19,600	698	280,546	10,654	7862	276	172	617	—	—
1974	31,124	977	19,300	687	221,363	8256	3082	108	231	833	—	—
1975	31,976	1004	20,000	712	248,139	9462	5429	191	229	824	—	—
1976	35,021	1097	22,300	794	228,933	8662	5005	177	263	947	—	—
1977	36,509	1146	22,500	801	221,569	8385	3909	137	268	965	—	—
1978	37,146	1166	22,300	794	235,309	8939	3552	124	242	871	7832	196
1979	36,653	1151	23,000	819	245,976	9301	5838	204	239	860	17,000	425
1980	34,883	1095	24,000	854	272,671	10,351	5440	190	235	846	66,900	1673
1981	37,348	1173	24,000	854	272,595	10,367	8138	285	—	—	80,500	2013

[a] Large amount of kerosene was imported that year. Data obtained from AB Svenska Shell, Visby, Visby elverk and Gotlands kraftverk.

comparison to total refined oil imports of about 273,000 m^3 yr^{-1} (Table 2.3); Gotland's oil production is exported for refining. There is speculation that oil may be produced from underwater fields off the east coast. Since the sharp energy price increases of 1973–1974, the use of oil, which accounts for a major part of imported energy, has declined. The cement industry, which is a very large consumer of oil, influences this importation significantly; it has done so by introducing a more energy-efficient process for making cement. Furthermore, more coal will be used as an energy source in this industry. Although fossil fuel consumption has declined somewhat during the 1970s the use of electricity has continued to rise. The combined effect has been a decline in total energy consumption (see Tables 2.3 and 2.4 and Figures 2.14 and 2.15). However, if total energy use is expressed in oil equivalents the decline has not been very substantial.

In terms of actual energy (heat equivalents) the proportion of energy imported by type has changed somewhat during the 1970s (Table 2.3 and Figure 2.16). The main imports of oils no. 1 and 5 have declined as a percentage of total energy imported, while gasoline has increased. Electricity accounted for a much larger percentage in 1978 compared to 1972. The economic cost of energy has risen dramatically as prices, as well as taxes, climbed during this period (Table 2.5). It seems as though price hikes in the 1970s led to dramatic declines in energy consumption, at least in the short-run (Figure 2.15). The cost of energy has become an increasing monetary burden, with total monetary outlays rising significantly (Table 2.4). Whereas the total energy consumption in 1977 was slightly lower than in 1970, the monetary cost (in current prices) of this energy was about three times as great.

If the total consumption of fuels and electricity is compared to the natural energy flows, it can be seen that most of the natural flows are greater than the purchased energies by (in some cases) very large factors if no corrections are

Table 2.4. Total Energy Consumption in Oil Equivalents, Average Linearly Weighted Energy Prices, and Total Monetary Cost for Energy, 1970–1978[a]

Year	Energy Consumption (PJ/yr)	Average Weighted Price (10^{-6} Skr/kJ)	Total Cost (Million Skr)
1970	13.9	5.44	75.6
1971	14.0	6.86	96.0
1972	15.6	6.76	105.5
1973	14.7	6.77	99.5
1974	12.7	13.60	172.7
1975	14.1	11.00	155.1
1976	13.5	14.20	191.7
1977	13.7	16.60	227.4
1978	14.0	—	—

[a] Energy is expressed in oil equivalents by converting imported electricity to oil equivalents by dividing by 0.3 and adding to energy values of all other fuels.

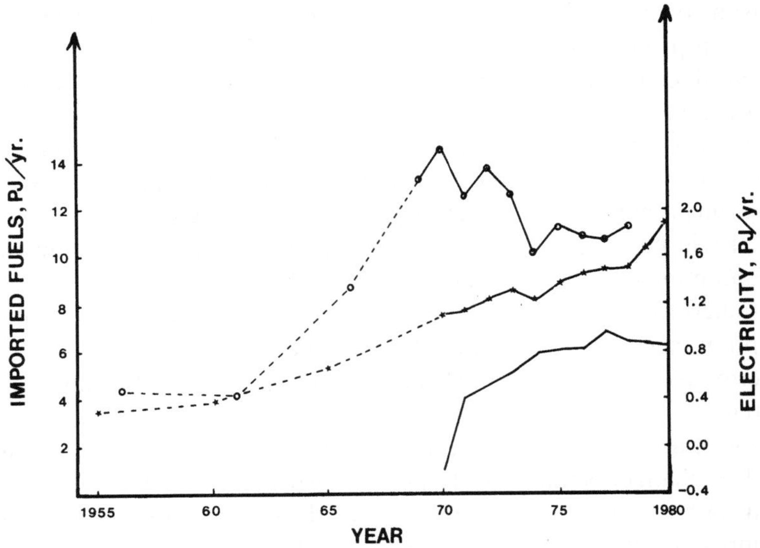

Figure 2.14. Fuel imports, electrical consumption, and imports of electricity for selected years, 1933–1980. (o, fuel imports; •, electricity consumption; –, net electrical imports).

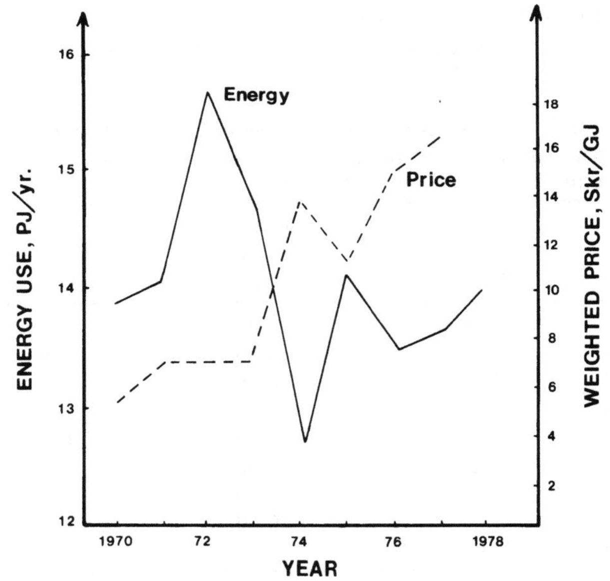

Figure 2.15. Total end-use energy consumption expressed in oil equivalents and weighted price for fuels and imported electricity (note; electricity expressed as oil equivalents).

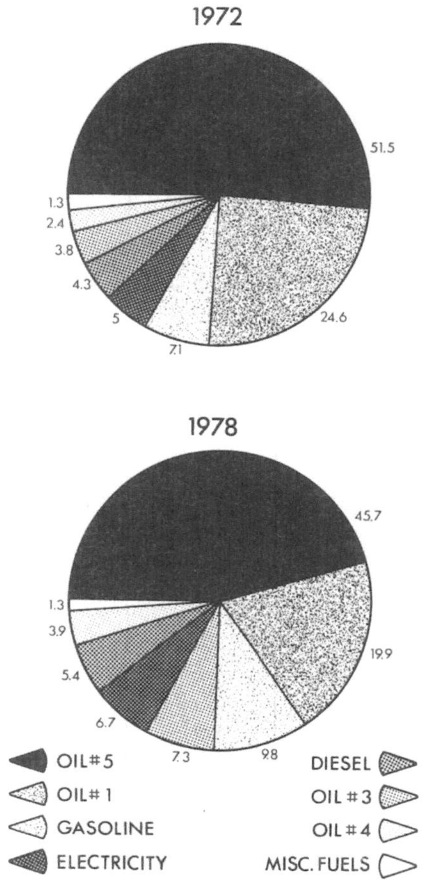

Figure 2.16. Pie diagrams indicating the change in the proportion of imported energy, 1972–1978.

made for quality differences. In particular, the work done by photosynthesis of about 100 PJ yr^{-1} is about six to seven times as great as the flow of about 14 PJ yr^{-1} of "cultural" energies—a ratio that indicates the degree of human activity in relation to the natural systems.

A more disaggregated description giving a perspective, at a glance, of the relative magnitudes of various energy flows of the region may be presented in a systems diagram. In our original monograph, which summarized results from the first years of the project, a large disaggregated diagram displayed and synthesized flows of energy, matter, money, labor, nutrients, and water to the subsectors in the natural and human systems (Jansson and Zucchetto 1978a, 1978b). Herein, we present a simplified energy systems diagram that gives a partial perspective on energy flows for 1972 (Figure 2.17). The calculations for this diagram are documented in our monograph; more details for each sector can be found later in the analysis of the subsectors. Data for electricity and

Table 2.5. Some Representative Prices for Energy on Gotland, 1969–1978

	1969	1970	1971	1972	1973	1974	1975	1976	1977	1978
Gasoline (öre/liter)	83.5	83.5	93.5	93.5	95.0	135.0	124.0	151.0	155.0	
Diesel (öre/liter)	61	60	75	75	75	57	43	59	58	
Oil No. 1 (Skr/m^3)	188	149	205	204	222	528	406	566	568	
Oil No. 5 (Skr/m^3)		64	121	121	109	354	289	319	450	475
Electricity (öre/kwh)						9.59	13.01	14.56	16.5	

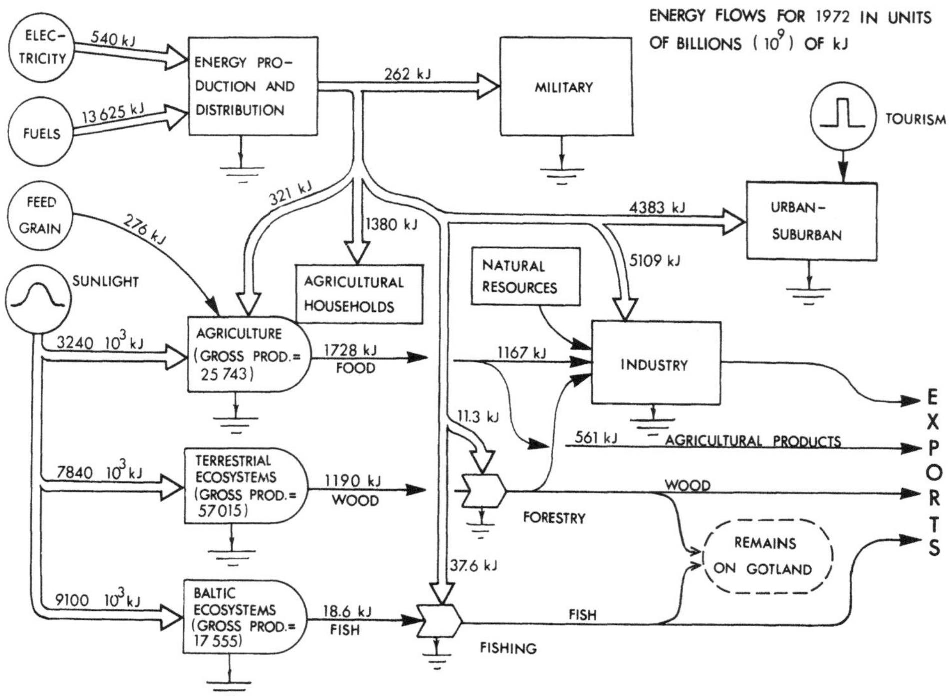

Figure 2.17. Distribution of energy for 1972 in units of billions of kJ to subsectors. Energy output for the energy production sector is the sum of fuels and electricity. Energy content of feed grain, food, wood, and fish are also indicated (Jansson and Zucchetto 1978a).

fuels were obtained from the power company and the fuel distributor. We used values from the ecologic literature of the various terrestrial ecosystems, as well as official statistics on harvests, to estimate productivities. Separate studies were conducted for agriculture, forestry, fisheries, and the military to estimate energy consumption. Although the renewable energy flows remain relatively constant from year to year, the flows in the economic system have changed considerably. Most of our detailed time series data for industrial activity is presented later in this chapter. These data allow comparisons of energy consumption in later years to that of the 5109 TJ in 1972.

Economic Overview

Gotland has a reasonable degree of economic diversification, considering its population and geographic size. Being an island, it is heavily dependent on imported resources and goods, as well as on external markets, for its exports. The internal resource base—which essentially consists of the use of solar energy in the form of biomass produced by photosynthetic systems, fresh water supplied by the hydrologic cycle, and the exploitation of geologic deposits—leads to extensive economic activity in agriculture, forestry, quarrying, and cement

Table 2.6. Percent of Total Employment on Gotland by Activity, 1972–1977

Activity	1972	1973	1974	1975	1976	1977
Agriculture	18.7	17.7	17.9	17.6	18.5	17.0
Quarries	0.7	0.7	0.6	0.6	0.6	0.6
Other industry	10.7	10.7	13.2	14.0	12.5	11.8
Electricity	0.8	0.8	0.4	0.4	0.7	0.8
Construction	8.1	8.4	5.7	6.1	6.9	7.9
Trade	13.0	12.5	12.9	14.5	13.9	13.6
Transportation	7.7	7.6	6.5	6.9	6.2	6.4
Banking	4.5	3.6	4.2	3.4	2.7	2.6
Other services	7.3	6.0	9.1	4.6	5.4	5.7
Public sector	23.2	23.7	28.9	28.6	30.4	31.7
Total employment	24,600	24,900	26,300	26,200	26,000	26,500

production. As far as the local economy is concerned, agriculture, and its coupling to the food industry account for a higher percentage of employment and gross regional product (GRP) than for Sweden as a whole—a result of favorable conditions for cultivation on Gotland and an area of 1.5 ha capita^{-1} of arable land, compared to 0.4 ha capita^{-1} for Sweden as a whole. Agriculture accounts for about 11% of the GRP and the food industry accounts for about 7.5%. Fisheries constitute a rather small contribution to the GRP but it became our conviction during the course of the study that there was some room for expanding fish-processing capacity. Gotland's location in the Baltic Sea made it attractive, and the diminished activity of Swedish fishermen in the North Sea stimulated fisheries in the Baltic. In 1981, the fish-processing industry was expanded, creating 25 new jobs.

Cement production is one of the main industries, centered at Slite, that results in both extensive quarrying for limestone and a large proportion of the total energy consumption. Cement production capacity is about double what it was in 1972. However, a sharply decreasing demand in Sweden and cement imports

Table 2.7. Percentage Contribution to the Gross Regional Product by Activity, 1972–1977

Activity	1972	1973	1974	1975	1976	1977
Agriculture	10.0	8.9	11.6	7.5	7.9	7.6
Quarries	2.0	2.0	1.6	1.6	1.6	1.5
Manufacturing	13.0	16.0	19.4	22.9	23.9	23.0
Food industry	4.0	4.3	5.5	4.3	4.6	4.4
Stone and soil	3.4	3.6	3.6	3.9	5.7	5.2
Workshops	4.4	6.3	8.1	12.0	11.1	11.1
Electrical	0.9	0.8	0.5	0.7	0.9	0.9
Construction	8.2	9.5	5.2	5.9	7.0	8.5
Trade	8.8	10.9	9.3	10.1	11.2	10.3
Transportation	5.6	6.3	5.1	4.9	4.4	4.4
Private services	10.0	11.0	9.1	9.1	7.7	7.0
Public sector	29.0	25.9	27.7	27.5	27.5	25.3

from Poland have resulted in production levels far below capacity. In terms of GRP, the stone and soil sector as well as quarrying activities make up about 5.5% and 1.5%, respectively, although these proportions change from year to year. Other important industrial activities consist of various kinds of manufacturing workshops, especially electrical workshops, with the firm of L.M. Ericsson being a major employer and generator of value added. However, in the last few years, L.M. Ericsson has scaled down its activities and eliminated hundreds of jobs. Important activities include construction, wholesale and retail trade, banking, and government (Tables 2.6 and 2.7). Public sector jobs account for about 30% of employment. Agriculture, trade, and manufacturing workshops are also important employment sources. Another major source of employment and income is tied to the tourist industry. It was estimated that tourism accounted for an economic output of about 80 MSkr and the direct employment of 2100 people in 1975.

In terms of quantity, imports are dominated by petroleum products, fertilizers, animal feed products, and food—although all sectors import goods to one extent or another. By weight, exports are dominated by stone, cement, wood, and food; but, again, many sectors produce goods for export. The flows of money to and from the central government of Sweden are important components of the economic picture that are strongly affected by Swedish regional economic policies. Decisions by the central government about locational or transportation subsidies help to encourage firms to relocate on Gotland. Decisions to locate government offices, such as Lottery headquarters to Visby, obviously have significant influences on employment. Additionally, government-sponsored activities such as military training and bases generate revenue for the local economy. Direct payments by the central government to Gotland occur in the form of tax-equalizing and operational subsidies to the municipality, as well as social security transfer payments (e.g., pensions and subsidies for children). On the other hand, taxes are paid to the central government, but these were lower in 1972 than transfer payments, resulting in a net subsidy from the central government of about 120 million Swedish Kronor.

The GRP is thus generated by many different private and public activities. For 1972, the percent contributions were somewhat as follows: (1) private consumption, 55.5%; (2) commune consumption, 22.5%; (3) military expenditures, 15.5%; (4) private investment, 12.6%; (5) commune investment, 2.2%; (6) housing investment, 3.3%; (7) exports, 41%; and (8) imports, −51.6%. The GRP has varied from 1972 to 1979, but there seems to have been somewhat of a decline over this period in both total and per capita value (see Table 2.8). It should also be noted that the percent of GRP contributed by workshop industries and the stone and soil sector has increased rather significantly, while agriculture, quarries, banking, and the public sector have somewhat declined.

The relationship between energy consumption and economic production is illustrated by plotting the ratio of direct energy use (in oil equivalents) to GRP at constant 1975 prices (Figure 2.18). One also could plot energy against GRP as discussed in Chapter 1. The direct energy use-to-gross regional product ratio (E/GRP) declined from 1972 to 1977. A large drop occurred in 1974 as a result

Table 2.8. Energy Consumption, Gross Regional Production, and Energy-to-Gross Regional Product Ratios, 1972–1977

Year	Gross Regional Product (Millions of Current Skr)	Gross Regional Product at 1975 Prices (Millions of Skr)	GRP/capita at 1975 Prices (in Skr)	Energy Consumption[a] (per capita, TJ)	Energy-to-GRP ratio at 1975 Prices (kJ/Skr)
1972	1158	1503	27,833	0.29	10,380
1973	1219	1469	27,229	0.27	10,007
1974	1512	1661	30,661	0.23	7646
1975	1555.4	1555.4	28,592	0.26	9071
1976	1681	1529	27,993	0.25	8829
1977	1789	1454	26,450	0.25	9422
1978	1833	1358	—	—	10,309
1979	2104	1451	—	—	10,177

[a] Energy consumption is in oil equivalents per capita (see Table 2.4).

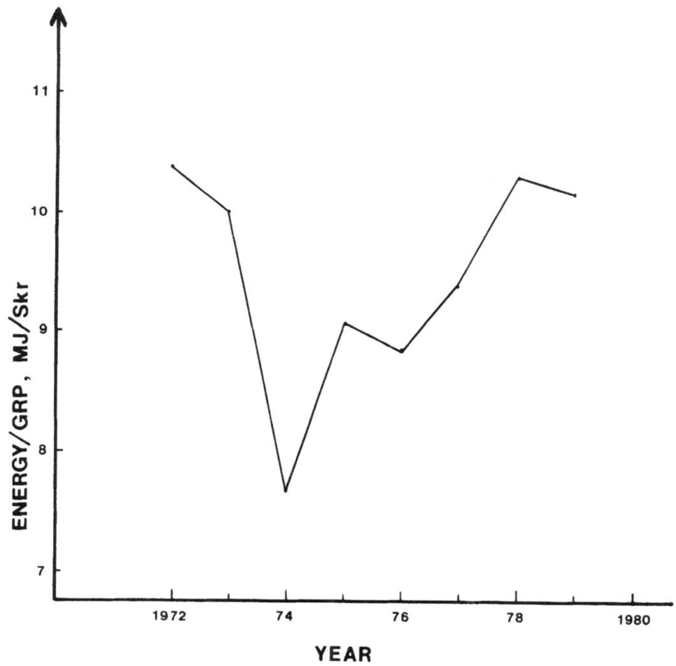

Figure 2.18. Energy-to-gross regional product ratio, 1972–1977. The GRP is expressed at 1975 prices (see Table 2.8) The energy is expressed in oil equivalents.

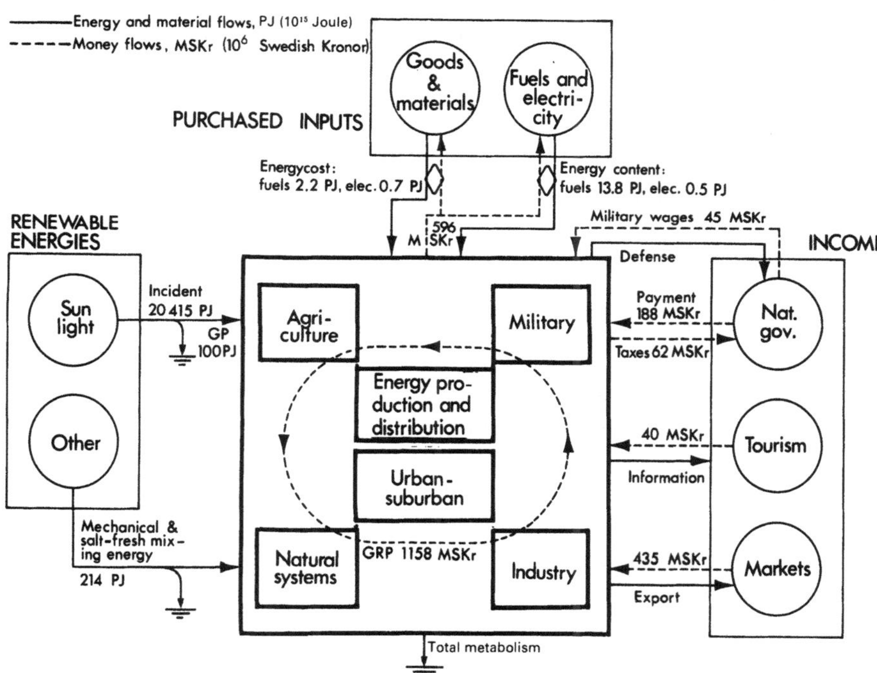

Figure 2.19. Macroview of Gotland's economy showing major money and energy exchanges with the outside world. The energy cost of imported goods and services is embodied energy value. The energy flows are in PJ, and the money flows are in millions of Swedish kronor (Data for 1972, Jansson and Zucchetto 1978a).

of oil shortages, price rises, and rationing connected to the world oil crisis. These events resulted in a decrease in energy for private use as well as in industrial processes. However, the ratio is seen to increase again from 1974 to 1978, implying somewhat of a return to "business as usual"; that is, it appears that energy and economic output remain rather strongly coupled if one compares 1972 to 1979. One other noteworthy important economic trend is the increasing percentage of employment in the public sector, whereas most other sectors have remained almost constant over the period 1972–1977. A macroscopic overview of main energy flows and monetary exchanges for 1972 is presented in Figure 2.19. In this figure, we simply indicate the total flows of energy, both renewable and purchased, to Gotland as well as the major monetary measures such as GRP, imports and exports, and transfer payments. Note that we have also included the embodied energy of imported goods; this is the energy expended outside Gotland to produce these goods (Jansson and Zucchetto 1978a). The indirect fuel support amounts to 16% of the fuel imports on the island (2.2/13.8), while the indirect electrical costs of 0.7 PJ are greater than the electrical imports of 0.5 PJ.

Resource and Environmental Issues

Although much of the Gotland study focused on energy resources and their relationship to economic activity and the environment, our general broad concern with energy processes in both the systems of humans and nature naturally led to a consideration of the system interactions of other kinds of resources. During our investigations, other critical resource assessments emerged that were equally important to consider for Gotland. Our approach, in deciding what resources to focus on, was to try to identify those resources that seemed to be most limiting to economic or ecosystem activity. Concerning the environmental aspect of regions, assessments had to be made as to which activities could generate the most impact. Resource and environmental impacts can also be thought of in energy terms, although we have not stressed this as part of our study. High-quality and concentrated ores, for example, need less energy input than low-quality ores to produce refined material; they have a higher free energy with respect to the environment. Limiting factor resources can also be thought of in terms of their value in allowing work to be done. On Gotland, water is limiting to agriculture; higher than average rainfall usually produces significant increases in photosynthetic production.

Water Quantity

Availability of sufficient water supplies is one resource area of great concern. This may be a general limiting factor to consider for most island systems. Consider the fluctuations over 1 year that, on the average, occur for precipitation and potential evapotranspiration (Figure 2.20). As would be expected, potential evapotranspiration is highest during the summer months because of elevated solar insolation and temperature, but rainfall is lowest so that an extreme hy-

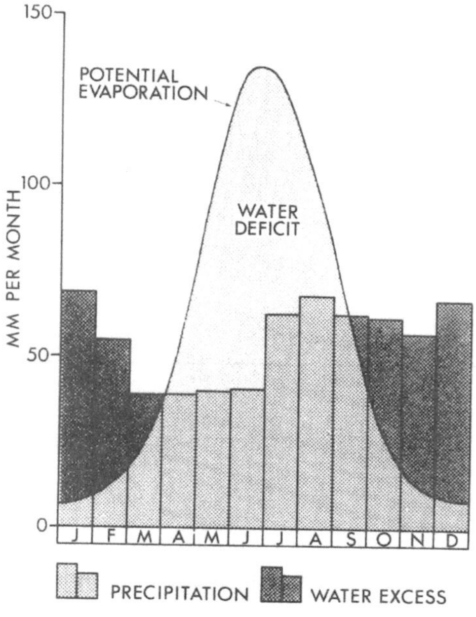

Figure 2.20. Thirty-year average (1931–1960) of precipitation and potential evapotranspiration over the year (Anonymous 1981b).

drologic deficit occurs during most of the summer. Analysis of the water balance also indicates that in June to August, low soil moisture conditions prevail that limit photosynthetic productivity in natural ecosystems, agriculture, and forests. In agriculture, for example, a very wet year in 1974 with total precipitation of about 740 mm, in comparison to average annual rainfall of 525 mm, produced agricultural harvests almost 30% greater than in more normal years (Jansson and Zucchetto 1978a).

Water may also be constrained in terms of supply for other human related activities. In the short-term, water shortages may not be evident, because surface water and groundwater storages can be exploited. However, in the long-term, these storages must be replenished at the rate at which water is lost or taken from them in order to maintain a sustainable system of humans and nature on the island. There are not extensive supplies of surface water, (basically, only two large lakes), and groundwater is lost from many areas of the island because of the limestone geology and its fractured nature. The historic drainage of wetlands has reduced groundwater recharge, increased runoff, and decreased water storage capacity. Again, because of the mismatch between potential evapotranspiration and precipitation during the summer months, supplies of water during this period will, of necessity, be constrained. However, it is during this period that the large tourist influx, now about 300,000 visitors per year, occurs. It exerts a large demand, in relation to the resident population, for water and other resources. For example, the water pumped by Visby waterworks exhibits a marked seasonality due to the increased demand that occurs during the

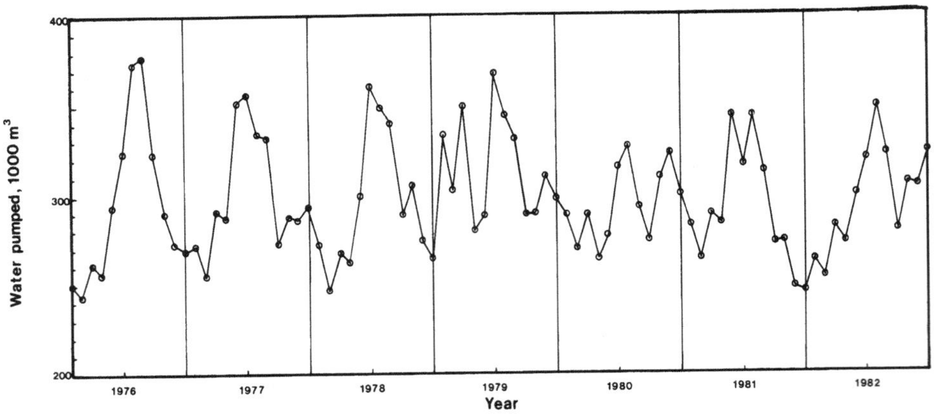

Figure 2.21. Water pumped by Visby waterworks, 1976–1982, exhibits marked seasonal behavior.

summer months (Figure 2.21). Recent events exemplify the effect of decreased storage in natural lands. During the dry year of 1983, creeks dried up and fish were killed or unable to migrate up to their spawning areas.

If additional water supplies are required, then exploitation of surface waters in the northern part of the island may be necessary, but at a cost. Further drainage of the small area of remaining wetlands would only aggravate the situation. Since water is a limiting factor, productivity of the terrestrial systems (especially in agriculture) could be enhanced through irrigation, which would generate economic benefits. However, this strategy would be contingent upon an adequate availability of water for all activities on the island during the growing season. Because of the competition for water by different users, we attempted to estimate water consumption on the island and to incorporate this information into our economic models.

A flow diagram of water use on Gotland for 1972 is presented in Figure 2.22, in which estimates have been made of principal sources of water as well as consumption for several broad categories associated with human activity (Hilding 1982). Use of water by the natural communities on land can be attributed to evapotranspiration, and it constitutes a very large quantity in relation to human consumption. In fact, the large quantities of water associated with the natural environment in the form of precipitation, evapotranspiration, and potential groundwater recharge seem to imply that there is no critical threat to water availability for human consumption. However, evapotranspiration is lost from the system; the extent to which potential groundwater recharge becomes usable storages in surface and groundwater is not fully known. Much of this water is lost to the Baltic Sea and available surface water and groundwater of acceptable quality are limited. The water supply problem is aggravated by environmental impacts, which reduce water quality in certain parts of the island. Notice that household consumption consists of both surface water and groundwater, with one segment of households not served by water treatment facilities

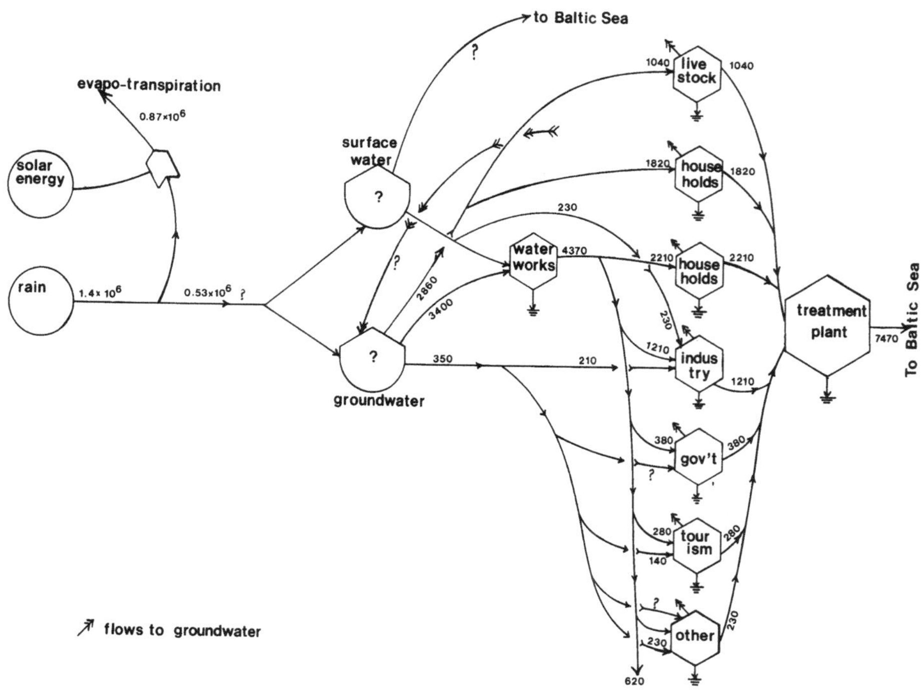

Figure 2.22. Water balance for the entire island of Gotland for 1972 in 1000 m^3/yr. (Hilding 1982).

and drawing water directly from groundwater. Other major demands include water for livestock as well as for industrial use. Flow models such as those presented for water (Figure 2.22) or energy (Figure 2.17) generate an overall system perspective as to quantities, relative magnitudes, and interdependencies. Regarding the combination of economic and environmental considerations into integrated models (as discussed later in Chapter 3), information on flows of materials to and from different economic sectors becomes essential to calculating resource requirements and pollutant emissions as a function of economic output. Limitations in the supply of some given input to economic activity would act as a constraint, as might happen with water in the future.

Water Quality

Human activity produces various kinds of pollutants, which are emitted to the environment and, depending on the quantity and type of pollution, affect it in numerous ways. Examination of the environmental situation on Gotland pointed to water quality problems, in the form of nitrate pollution, as an important area of study. High levels of nitrate intake by humans can lead to methaemoglobinemia in infants or to the formation of carcinogenic nitrosamines in humans and animals. The average nitrate concentration in drinking water on Gotland was 8.8 mg liter^{-1} in 1972, compared to 2 mg liter^{-1} for Sweden. However, these

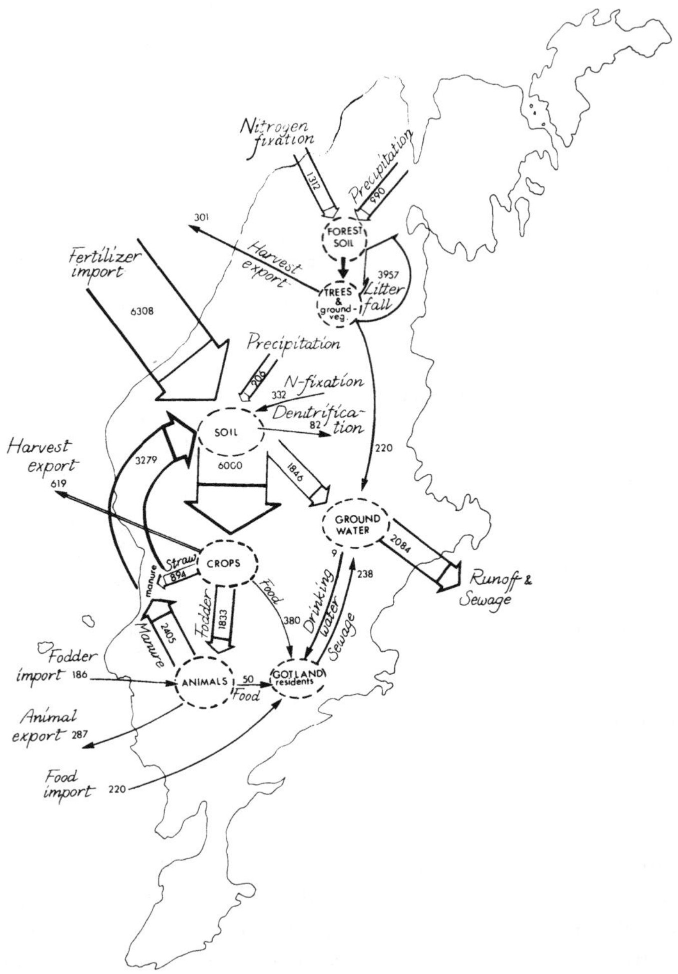

Figure 2.23. Nitrogen budget for the principal systems of agriculture and forestry in 1972. Note the major input from fertilizers. All values are in metric tons per year (see Appendix I) (Jansson and Zucchetto 1978a).

average statistics disguise the fact that nitrate concentrations in the groundwater—especially in agricultural districts, around the town of Visby, and near drained bogs—registered levels near or above 100 mg liter^{-1}. These levels greatly exceed the permissible limit of 50 mg liter^{-1} for children under 1 year of age. To gain a quantitative perspective on the relative magnitudes of nitrogen flows in the region, nitrogen budgets were constructed for the entire island (Figure 2.23) as well as for the site at Lummelunda north of Visby. The influence of human activity in agriculture, where fertilization is about 70 kg nitrogen ha^{-1} yr^{-1}, has greatly modified the natural nitrogen budget. The fertilizer and manure nitrogen flows are much greater than for those associated with nitrogen fixation or precipitation. These large flows lead to infiltration of nitrate into groundwater,

Table 2.9. Estimates of Biological Oxygen Demand for Different Economic Sectors, 1972[a]

Code	Category	BOD_7 (tons/yr)
1.1	Agriculture	2830
1.2	Forestry	—
2	Quarries	1
3.1	Food industry	230
3.2	Textile industry	—
3.4.1	Wood industry	1
3.4.3	Graphics industry	1
3.5	Chemical industry	1
3.6	Stone and soil	3
3.8	Workshops	5
4	Electrical and district heating	1
5	Construction	12
6	Retail, restaurants, and hotels	48
7	Transportation	7
8	Banks and insurance companies	3
9	Government, hospitals, schools, and military	80
	Households	1060
9.2.0.1	Water treatment	1666
	Tourism	85

[a] Hilding (1982).

as well as runoff to streams and the Baltic coastal systems. They also seem to have accelerated denitrification processes and nitrogen release to the atmosphere. Water quality models were used to estimate nitrate impacts as a function of fertilization. Also, the impacts on the coastal ecosystem were investigated, to some extent, in the fishery models. Results of these modeling efforts are presented in Chapter 3.

From what information we have been able to gather, problems of chemically toxic or radioactive materials are not present to any significant degree and they do not constitute an aspect of any water quality problems. However, many wells on Gotland contain water that is too saline for human consumption. One other major source of water pollution consists of the production of organic material, which exerts a biochemical oxygen demand in aquatic systems. In the past, especially because of the presence of the food industry, organic loading was a serious problem in certain areas such as Visby. The building of sewage treatment facilities in the 1970s greatly reduced organic matter outlets, but biological oxygen demand (BOD_7) emitted to the environment is still extensive. Estimates are presented in Table 2.9 for different economic sectors; they show agriculture, food industries, households, and sewage treatment as the major contributors to organic loading. Again, information is presented by economic sector, because economic and ecologic models can then be interfaced with this information to couple levels of economic activity with environmental impact.

Other Environmental Impacts

Various kinds of land-use and economic activity generate a diverse number of environmental and ecologic effects that differ in their quantitative and qualitative extent. Attention has already focused on the historic changes in the environment, especially with regard to land-use change, that have occurred as well as on impacts on water quality and quantity, water quality distinctly related to agricultural land use, and types of cultivation. Although extensive deforestation was a threatening environmental problem in the nineteenth century—leading to a series of environmental impacts such as erosion and decline of water supply—decreasing demand for firewood made possible a regrowth of the forests. At present, the threat of excessive deforestation is practically nonexistent. Several changes in land use are being generated by different activities; individually, these seem small in comparison to the whole system, but locally they may present severe issues. For example, the extensive expansion of the cement industry in Slite will result in extensive quarrying in the area of File Hajdar and will lead to the eventual destruction of a natural heath ecosystem of high biological diversity. Various forms of quarrying around the island increase hydrologic runoff and affect water recharge to groundwater supplies. Although the resident population has not changed significantly in absolute numbers, increases in urbanization and the standard of living have had their impacts. For instance, natural and agricultural land around Visby has been converted into residential and commercial areas. Furthermore, the large increase in the number of tourists has led to construction of summer homes, hotels, and campsites, as well as to an increased number of visitors on the beaches and other natural lands. It is also known that intensive farming of the drained wetlands has resulted in extensive erosion and decline of the peat resources (Nilsson 1982). Increased runoff, activities in parts of the coastal system, and fisheries all generate impacts on the Baltic coastal system. One activity whose impact is difficult to evaluate because of secrecy restrictions is the military, which engages in extensive maneuvers and exercises on heathland and gravel beaches, particularly in the northeastern part of the island. Soldiers, vehicles, weapons, bombing runs, and construction all take their toll. Because of high winds and relatively low population densities, air pollution does not appear to be a significant problem, except in Visby and some other local areas where automobile traffic is heavy and emissions from industries and power plants are concentrated. However, erosion occurs as a result of high winds. Problems of acidification from acid rain, which is a major issue in Sweden, is not important on Gotland because of the limestone geology.

Renewable Energy Technologies

The Swedish debate concerning energy for the future has seriously considered the potential of renewable energy technologies to support a "solar" Sweden (Lönnroth et al. 1977) as an alternative to the use of nuclear power. The Swedish electorate voted in 1981 to complete nuclear power plants that were already under construction or planned, which will eventually result in 12 operating nuclear power plants. However, it was also decided that these power plants would be phased out at the end of their useful lives and that alternative technologies would be substituted instead. In the spirit of an eventual switch to renewable technologies for Sweden, our study has attempted to make some assessment of the potential contribution from locally available renewable energy sources; in Chapter 3, several optimization models are considered for determining an appropriate mix of energy technologies. Several renewable energy supply systems may have a reasonable chance of succeeding on Gotland because of relatively low population densities, a relatively high number of hours of sunshine, considerable wind energies, and a long coastline with high wave energy. In addition, extensive agricultural and forest lands may allow economically attractive harvests of straw, wood waste, and biomass plantations. On Gotland, however, enthusiasm about biomass "plantations" must be tempered with the reality of water constraints. Wind energy is high on the island (an average wind velocity of 6 m sec^{-1} during the year amounting to 150 PJ yr^{-1} within 100 m of the ground); part of this could be converted to electricity with aerogenerators. A large 2 MW aerogenerator has been constructed and experiments have been conducted during the early 1980s. Another proposal is to harness wave energy by means of system of buoys off the coast, each of which would generate 0.45 TJ yr^{-1} of electrical energy. Energy plantations could be situated on small areas of land that are not being used for agriculture or forestry or that are of critical environmental concern (estimates for such areas include 340 ha of marsh, 500 ha of forest, 68 ha of deciduous glade and 474 ha of grazing land). Also, one may let energy plantations compete with food crops for agricultural land, since there is an excess production of crops on Gotland. Other organic energy could be derived from wood waste, straw, dung, and garbage. Solar energy, as well as geothermal sources, could be used for heating houses. Heat pumps using the temperature difference of waste water and Baltic seawater have been installed in Visby for district heating. Estimates for the potential from renewable energy technologies are summarized in Table 2.10, and these should be compared to consumption levels for 1972 illustrated in Table 2.11 (Zucchetto and Jansson 1981, Ahlbom 1982).

Potential contributions from renewable energy technologies appear to be high; however, questions of net yield, net energy, and economics arise. Several studies of renewable energy technologies (Kjellström and Gustafsson 1976, Anonymous 1977b, Anonymous 1979c, Eneroth et al. 1979), and work during our project, indicate energy yield ratios greater than 1, although ratios varied depending on the assumptions made. Energy plantations included the following costs: fertilizer production, transportation and spreading, irrigation, establishment of seedlings, and harvesting. Biomass production, depending on the type of soil, resulted in a yield ratio as high as 100 with no fertilization or irrigation and as low as 2.2 with intensive applications. Calculations for 2-MW wind-electrical generators, assuming a cost of 12 öre kWh^{-1} and an embodied energy cost of goods of 10,000 kJ Skr^{-1}, resulted in yield ratios of about 10. Wave-electrical buoys, assuming a lifetime of 20 years and an embodied energy cost of 10,000 kJ Skr^{-1},

Table 2.10. Potential Contributions from Renewable Energy Technologies for Gotland (TJ/yr)

Technology	Fuels	Electricity
Wind-electrical[a] (100 generators)	—	1300
Wood wastes[b]	200–1000	—
Straw (methane gas)[c]	830	—
Dung (methane gas)	290	—
Garbage (methane gas)	150	—
Wave-electrical (360 buoys)	—	145
Solar heating[d]	175	—
Totals	1645–2445	1445

[a] Two MW generators; 200 probably could be placed on the island.
[b] Only using one-third of current cutting would give the low figure.
[c] Two tons (dry matter)/ha for 82,000 ha.
[d] Both old and new houses.

Table 2.11. Levels of Selected End-Use Energy Requirements for Gotland (TJ/yr), 1972

Activity	Fuels	Electricity
Industry		
Stone and soil	4000	1160
Food	290	146
Quarries	12	110
Workshops	7	87
Graphics	6	22
Wood	4	14
Chemical	1	16
Textiles	1	0.4
Agriculture	270	52
Agricultural households	1300	63
Forestry	9	—
Urban-suburban	2500	590
Totals	8400	2260

resulted in an energy yield ratio of about 2.5. Biomass, wind-electrical and wave-electrical technologies, although not having energy yield ratios as high as fossil fuels—which can range from 6 to as high as 40 (Odum et al. 1976)—seem promising from this perspective. Best estimates on costs per unit of energy by several investigators also indicate attractive prices (see Table 3.18).

Energy/Economic Analysis of Subsectors in the Region

This section contains some of the empiric results of analyses that were conducted for various activities or subsectors within the region of Gotland. In essence, this is data analysis of activities within the region with a focus on defining and measuring parameters associated with energy and economic requirements. Energy-to-economic measures have been calculated wherever possible. Some of these analyses are contained in a previous publication about Gotland (Jansson and Zucchetto 1978a). All of the details are not repeated here, but rather a summary of the main results as well as the inclusion of any new updated results are discussed. The subsectors considered are agriculture, several ecosystems, fisheries, forestry, industry, the military, tourism, transportation, and the urban sectors. Those readers interested in energy and economic analysis will find this information of interest not only as a characterization of the activities on Gotland, but as a basis of comparison for similar activities in other areas. Furthermore, the detailed quantitative study of these sectors prepares the way for system models that either represent individual subsectors of the region or characterize the interaction among the subsectors of the entire island. We do not document all of the calculations, since these can be found in the cited publications. There is nothing highly untested or theoretic, such as energy quality factors being used in these calculations. We rely on actual amounts of energy consumed from statistical sources, from the planning authorities, or from personal interviews in some cases; fuels and electricity are converted to energy values through heat value conversion factors. Likewise, economic estimates are gathered from statistical sources. Productivity in the ecosystems was based on the best field estimates and experimental data available either for conditions on Gotland or for similar sites on the mainland. Following from an energy basis viewpoint, those systems dependent to a high degree on solar energy are presented first.

Natural Systems

The "natural" or photosynthetic systems on Gotland have already been described in some detail. The focus in this section is on the energy flows associated with these systems. Since the landscape has been modified by human interference over periods of time, the word "natural" may be somewhat misleading; however, these assemblages of plants and animals are usually referred to as ecosystems. The photosynthetic productivity of the island's ecosystems has always been the basis for the economy of the human population. Even now, during a period of intense use of fossil fuels and electricity, systems based on photosynthesis contribute substantially to the economy. Only a small percentage of the enormous amount of solar energy that falls over the natural systems is incorporated as chemical energy in plant biomass, which supports the growth, maintenance, and reproduction of the components of the various ecosystems. This photosynthetic fixation provides energy for material flows, and it results in products that are directly useful for the human economy, such as wood for building or burning or food for human consumption. The maintenance of a diversity of flora and fauna contributes to recreation, aesthetic appreciation, or hunting. The human economy also exploits the work that photosynthesis performed in the past by using rich peat soils for productive agriculture; the soils are a nonrenewable resource if not carefully husbanded. The natural systems that have been considered are marketable coniferous forests, mixed deciduous

Table 2.12. Energy Flow in Gotland's Ecosystems, 1700 and 1980

	1700			1980		
	Area[a] (km)	Gross[b] per km^2 (TJ)	Gross Total (PJ)	Area[c] (km^2)	Gross[b] per km^2 (TJ)	Gross Total (PJ)
Forests	1785		47			
Coniferous	(1460)	30	43.8	1310	30	39.3
Forested heath	(325)	10	3.2	139	10	1.4
Wooded pastures	448	29	12.9	203	29	5.9
Mixed deciduous woods				209	38	8.0
Meadows	410[a]		7.9			
Actual meadows	(310)	23	7.1	3	23	0.1
Beach meadows	(~25)	23	0.6	26	23	0.6
Beach, other types	~25	—	—	36	—	—
Heath without trees	(50)	4	0.2	268	4	1.1
Mires	236	69	16.3	31	69	2.1
Lakes	47	12.6	0.6	25	12.6	0.3
Agriculture	195	10	1.9	830	29.6	24.6
Urban areas	7	—	—	10	—	—
Sum			86.6			83.4

[a] From Lindquist's (unpublished) analysis of the taxation map of Gotland, 1700 A.D. Areas in parenthesis are estimated values.
[b] Gross production per km^2 based on calculations presented by Jansson and Zucchetto (1978a). Gross production in agriculture in 1700 is assumed to be one-third of the present. Gross production of meadows is assumed to be the same as for beach meadows.
[c] Areas from Jansson and Zucchetto (1978a) are corrected for changes in agriculture and coniferous forests.

forests, heathland with trees, grassland heath, wooded pastures, mires, lakes, beaches, and the Baltic coastal system. Agriculture can be considered as an intensely managed photosynthetic ecosystem. Estimates of the productivities of these systems, together with agriculture and urban areas, are summarized in Table 2.12. Figure 2.24 highlights the enormous amounts of energy associated with the photosynthetic process. Total photosynthetic production depends on the areal extent, as well as on the intensity of production. It is dominated by coniferous forests, agriculture, and coastal ecosystems, with the other systems contributing in a declining scale. The energy intensity graph (Figure 2.24B) shows the importance of water and soil interactions; mires are the most productive systems; with production intensities declining as systems with less water or thinner soils are considered. Note that on a photosynthetic basis, agriculture does not perform much better than some of the other ecosystems, and it is lower than mires, deciduous forests, and coniferous forests. Of course, the productivity of agriculture is more directly channeled into useful food products, even if the agroecosystem has not improved on nature in terms of the ability to capture photons. It is also interesting to note the much lower productivity of the Baltic coastal systems in comparison to the terrestrial systems.

As explained later in the "Forestry" section in this chapter, the primary production of the exploited forest ecosystems supports tree and understory vegetation, of which 1190 TJ were harvested in 1972 by private individuals and

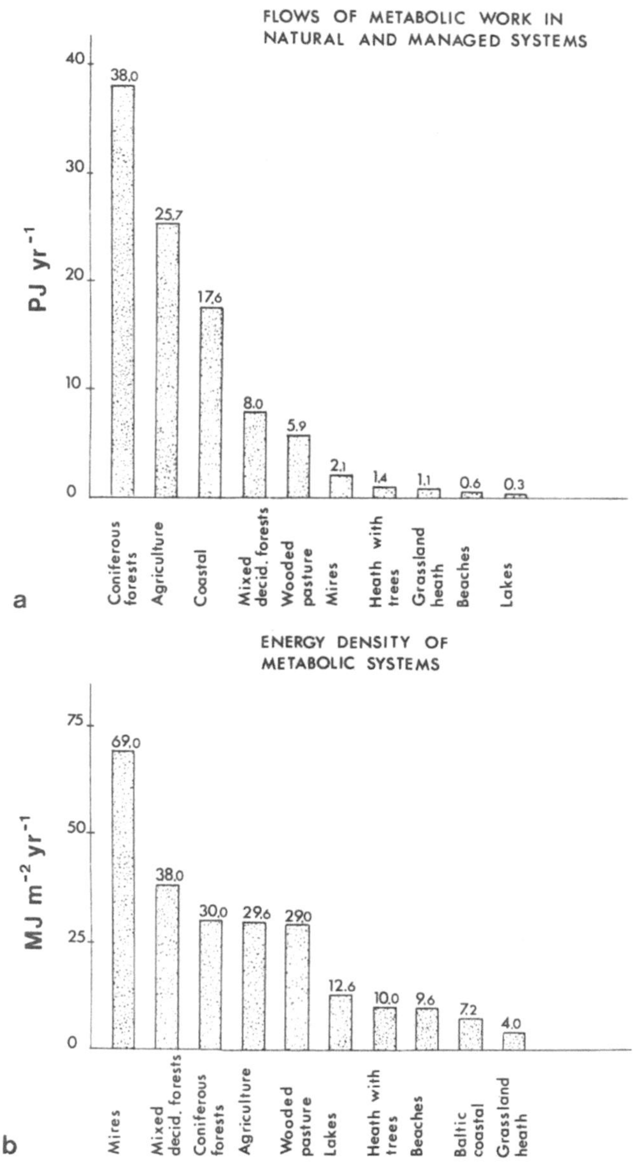

Figure 2.24. (a) Total gross photosynthesis for different ecosystems on Gotland as of 1972 in PJ yr^{-1}. (b) Gross photosynthetic intensity (MJ m^{-2} yr^{-1}) for different ecosystems (see Appendix I) (Jansson and Zucchetto 1978a).

companies. Deciduous woods are relatively rare; management of some of these areas for grazing in so-called "ängen" during previous times channeled energy into food production as well as wood biomass. The heathland systems are fairly low in productivity because of thin soils and wide ranges in temperature and moisture availability. These areas are considered to be almost worthless for wood production since trees are slow growing and biomass is often as low as

15 m^3 ha^{-1} compared to an average of 83 m^3 ha^{-1} in the commercial forests. These areas can support sheep grazing. Our estimates suggest that on existing heathland a population of 40,000 ewes could be supported, assuming one ewe per ha^{-1}. Moreover, heathlands have a rich diversity of flora that supports a food chain of rabbit, hares, and foxes; hunting of these species occurs throughout the island. Indirectly, the presence of the heath vegetation retards erosion, acts as a wind shelter, and regulates water flow, all of which contribute to the human system by creating a more favorable environment.

Mires have been reduced significantly in area over the past 150 years. The extensive depositions of peat that they have generated have contributed to the success of these areas, which were drained for agriculture. Wetland areas were used in the past for fishing, hay harvesting, cattle and sheep grazing, and harvesting for roofing material, whereas their presence heightened groundwater tables and stored water during periods of drought. The energy content of the peat is high, approximately 23 kJ g^{-1} (Post 1929), and the layers of peat in the mires constitutes a storage of chemical potential energy of about 6900 PJ. Lakes are a small part of the overall island system, but they are important because of their role as existing or potential sources of drinking and irrigation water. Average gross primary production is about 300 g carbon m^{-2} yr^{-1}; the production of edible freshwater fish amounts to only 10 kg ha^{-1} yr^{-1} for a total of 0.1 TJ yr^{-1}. The main types of beaches consist of about 6.5 km^2 of rocky coast, 22.2 km^2 of gravel beach, 7.5 km^2 of sandy beach, and 26.2 km^2 of beach meadows for a total of 62.4 km^2. The photosynthetic production of these areas is small; but on the more protected beach meadows, great masses of fertilizing algae accumulate, resulting in a production high enough to support cattle, sheep, rabbits, and birds (net production, about 5430 kg ha^{-1}). Finally, the coastal system constitutes a large area if depths down to 30 m are considered—the depth at which photosynthesis is almost zero. Gross primary production of the benthos to a depth of 30 m was estimated as 94,600 tons or 2120 TJ yr^{-1}, whereas gross photosynthetic production of phytoplankton was about 150 g carbon m^{-2} yr^{-1} for a total of 15,400 TJ yr^{-1} (Ackefors and Lindahl 1975, Ankar and Elmgren 1976, Jansson and Kautsky 1977, Jansson and Wulff 1977). This production supports populations of commercial and noncommercial sealife as well as seals and seabirds. Large populations of birds congregating around the island have been estimated to consume about 11.3 TJ yr^{-1} of bottom animals and 16.6 TJ yr^{-1} of fish. The consumption of fish by birds is about as large as what we estimated for Gotland's fisheries, which amounted to 18.6 TJ in 1973 (Jansson and Zucchetto 1978a). The present population of gray seals (roughly 50) in the waters around Gotland is small compared to the estimated 1000 seals that existed in the 1930s, when seal hunting was an important economic activity. Fish consumption by seals is estimated as 0.6 TJ yr^{-1} (Söderberg 1977). Part of the algal production in the coastal waters is washed ashore; this constitutes a nutrient and energy input to the terrestrial system, much of which was used as fertilizer in earlier times. A conservative estimate of this transfer from the coastal to the terrestrial systems is 6.7 TJ yr^{-1} or about 11.3 tons nitrogen and 1 ton phosporus, whereas the flow from the land in terms of runoff and sewage outlets amounts to about 174 TJ yr^{-1}, or 2300 tons nitrogen and 330 tons phosphorus.

Estimates of productivity were also made for preindustrial times in the year 1700 and were compared to conditions in 1980 (Table 2.12). Total area and production for most of the natural systems declined, whereas agriculture expanded by an order of magnitude; however, total photosynthesis has remained almost the same, with a decrease of 3.7% from 1700 to 1980. The great changes have occurred in agriculture. By virtue of this expansion, ecosystems in wetlands and meadows have been extensively displaced.

Agricultural System

Extensive elaboration of the calculations, assumptions, and data associated with the agricultural system can be found in Jansson and Zucchetto (1978a) and Zucchetto and Jansson (1979). Agricultural activity on the island is of prime importance to the economy, and it contributes a substantial amount to the GRP. Since the end of World War II, the introduction of machines to replace human and animal power and the extensive use of fertilizer has enhanced production per unit of land substantially. The number of horses declined from about 15,000 to about 1000 from 1944 to the middle of the 1970s while tractors have reached a total of about 6000. This replacement of animal power by fossil fuels has allowed photosynthetic productivity to be channeled into human food products, rather than in support of work activities by horses. These data allowed us to approximate the replacement of horses by tractors in energy terms, with the use of average power factors for both. The main crops are sugar beets, grain, potatoes, and oil crops (rape), whereas the most important livestock consist of sheep, pigs, and cattle. Annual harvests for crops are about 13,000 tons of sugar beets, 50,000 tons of grain for fodder, 30,000 tons of grain, 15,000 tons of potatoes, and 10,000 tons of oil crops, which corresponds to about 1700 TJ yr^{-1}. Livestock populations amount to about 80,000 sheep, 55,000 cattle, and 55,000 pigs, with animal production amounting to about 500 TJ yr^{-1} (both edible and nonedible parts). Animal protein production was enough to supply the protein requirements of 180,000 people (70 g $person^{-1}$ day^{-1}). The total energy output of edible food produced would be sufficient for the energy requirements of 400,000 people (10,800 kJ $person^{-1}$ day^{-1}), which greatly exceeds the need of the resident population. These calculations were accomplished by converting to dry weight and using standard protein and energy content estimates for different crops.

The area under cultivation has remained steady at about 81,000 ha for many decades. However, other aspects are far from a steady-state, with rising inputs of fuels, machinery, feed, and fertilizers and declining inputs of labor. The application of technology and fertilizers has increased yields considerably. The total harvest has nearly doubled since the 1940s; fertilization increased from about 15,000 tons in 1946 to current levels of about 55,000 tons. This amounts to an intensity of fertilization of about 680 kg ha^{-1} yr^{-1} in comparison to values in 1946 of about 185 kg ha^{-1} yr^{-1}. We have also noted a diminishing returns effect in that yields cannot be substantially increased by higher levels of fertilization, at least in years of average precipitation. Water is apparently the limiting factor at this point in time in agricultural production.

Table 2.13. Summary of Energy and Economic Measures for the Agricultural System on Gotland, 1972[a]

Direct electricity cost	51.9 TJ or 155.7 TJ oil equivalents
Direct fuel costs	269 TJ
Total direct energy costs	425 TJ
Indirect energy costs[b]	983 TJ
Total energy costs	1408 TJ
Value added	157.5 million Skr
Value of sales	351 million Skr
Vegetable plus animal production	1730 TJ
Direct energy per unit of value added	2700 kJ Skr^{-1}
Total energy costs per value of sales	4012 kJ Skr^{-1}
Total sunlight	3240 PJ
Total energy costs per unit of sunlight	0.00043
Vegetable plus animal production per unit of sunlight	0.00052
Total energy costs per unit (of vegetable plus animal production)	0.84

[a] See notes in Appendix II.
[b] The sum of costs for fertilizer, capital, goods, grain, services, and feed. Energy is expressed in oil equivalents.

The major economic and energy parameters evaluated for the agricultural sector are summarized in Table 2.13. Direct fuel costs refer to diesel, fuel oil, and kerosene, of which 21% refers to indirect energy to extract, process and transport the fuels (Anonymous, 1975). Indirect energy costs are those embodied in goods and services. Sunlight was calculated from meteorologic measurements of insolation and total area. Economic data and harvests were calculated from

Table 2.14 Summary of Energy Costs for Sheep Production System, 1972[a]

Direct energy costs	12.45 TJ
Direct electrical costs for mixing feed	2.9 TJ or 9.67 TJ oil equivalent
Indirect energy costs[b]	27.9 TJ
Total energy costs	50.0 TJ
Food fed to sheep	282.0 TJ
Sheep production	31.5 TJ
Sunlight falling on cultivated and grazed areas	365,000 TJ
Total energy cost per unit of sheep production	1.6
Sheep production per unit of sunlight	8.63×10^{-5}
Trophic efficiency (sheep production/food input)	11.2%

[a] See notes in Appendix II.
[b] Indirect energy costs for tractors, commercial fertilizer, plant protection, seed, machines, food, medicine, and other goods.

agricultural statistics. The analysis of the agricultural system as a whole was conducted for 1972 as a representative year, with economic values presented in 1972 prices. Indirect energy costs for goods and services were about 130% higher than the direct energy costs, while the ratio of direct energy consumption to value added was significantly less than for the economy as a whole. As expected, the flow of sunlight was orders of magnitude higher than the other energy flows, although it is far more diffuse. The overall ratio of 0.84 of total fuel and electrical costs to that of food production from the agricultural sector compares quite favorably with many agricultural systems (Pimentel and Pimentel 1979). Furthermore, the agricultural system contributed about 13.6% to the GRP.

Some cursory analysis was conducted solely for sheep raising, since this is a well-known and important activity on Gotland. Examination of Table 2.14, which summarizes this analysis, shows that about 31.5 TJ of sheep meat and wool were produced. This required the expenditure of 12.5 TJ of fuels and 2.9 TJ of electrical energy for a total of about 21.2 TJ in fuel equivalents. Indirect energy costs amounted to 27.9 TJ for a total energy cost of 49 TJ in fossil fuel equivalents. The overall production required 1.6 units of energy input for each energy unit of sheep products produced, and the conversion efficiency of food to sheep products amounted to 11.2%.

Fisheries

Although fisheries represent a small part of the total economic picture in terms of sales, they are of interest because they exploit the natural resources of the Baltic Sea. Thus, a reasonably direct coupling exists between solar energy and fossil fuel energy required to harvest the fish species, which feed indirectly on marine photosynthetic production. Species of fish that dominate annual harvests depend on the specifics of age-class dynamics, which are dependent on environmental conditions in years prior to the catch; for the Baltic Sea, cod and herring are dominant. Although herring is usually the dominant species by weight in the harvests, there are times when cod is more abundant and is caught in larger quantities, such as in 1951 and 1979. In the early 1980s, cod continued to be dominant in some years. Figure 2.25 and Tables 2.15 and 2.16 give some historic views of the changes in the fish catch that have occurred since the late 1930s, as well as income associated with the different fish species. The income generated depends very much on price. For example, salmon, which are a small percentage of the catch by weight, represent a large percent of the income. The price of the fish seems to be inversely related to the position in the food chain; the higher up the food chain the less abundant, and higher quality fish bring a higher price.

As in agriculture, fisheries have had large capital investments in the form of big boats that use large amounts of diesel fuels, although there are still many small boats operated by fishermen. The introduction of modern technology and energy input has resulted in increased catches of about 5000 tons in 1979 compared to catches of about 2500 tons prior to 1940. In the same period, labor requirements have decreased, with a total of 244 workers in 1973 compared to

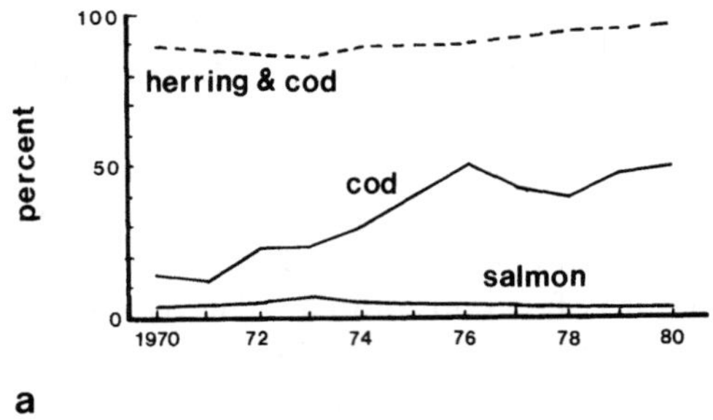

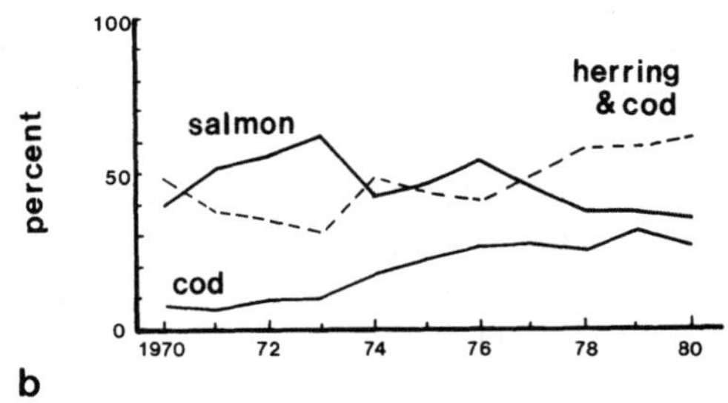

Figure 2.25. Percent of catch and income by species, 1970–1980. (a) Percent of total catch. (b) Percent of total income. (Limburg 1983).

about 1250 in 1920. The protein value of the fish catch can satisfy the protein demand of over 30,000 people, which is a significant percentage of the existing population of the island. Of course, much is exported, and cod is especially important for the frozen fish industry both in Sweden and abroad. One strategy, initiated by the Gotland planning authorities to increase employment, has been to expand the fish processing industry. The location of Gotland in the central Baltic fishing areas and large catches by Swedish fishermen justify the construction of new fish-processing facilities.

Analysis of the trophic energy flows in Baltic food chains can be found in previous publications (Jansson and Zucchetto 1978a, 1978b). Detailed analyses of the use of diesel and gasoline fuels for the catching of different species were undertaken for 1973 (Figure 2.26) as well as for herring and cod in 1979 (Figure 2.27 and Limburg 1983). The energy cost ratio is defined as the direct energy for operation (diesel fuel and gasoline) per unit of energy value of catch. Although a complete energy analysis should include the indirect energy costs for goods, data were not sufficiently disaggregated to derive this cost by species. Calculation also was made of the protein value of the catch and the direct energy cost per unit of protein energy (Figure 2.26B). Herring, flounder, and cod are quite similar in the energy required per unit of protein caught; on the other hand, salmon is very expensive in terms of energy. The total fisheries were also analyzed for 1973 as well as for other selected years (Table 2.17). The

Table 2.15. Catch of Different Fish Species on Gotland for Selected Years[a]

	Herring		Cod		Flounder		Salmon		Other		Total	
Year	Tons	TJ	Tons	TJ	Tons	TJ	Tons	TJ	Tons	TJ	Tons	TJ
1920	1418	9.5	488	1.70	102	0.30	34	0.30	49	0.20	2091	12.0
1944	1732	11.6	204	0.70	138	0.50	87	0.80	96	0.40	2257	14.0
1971	2234	14.9	173	0.58	119	0.44	132	1.19	256	1.08	2914	18.2
1972	1645	10.9	317	1.06	143	0.53	141	1.27	344	1.44	2590	15.2
1973	1850	12.4	575	1.95	162	0.60	259	2.30	316	1.30	3162	18.6
1974	2154	14.3	591	1.97	109	0.36	187	1.68	567	2.38	3608	20.7
1975	1650	11.0	965	3.22	119	0.44	160	1.44	470	1.97	3364	18.1
1976	1475	9.8	1528	5.10	140	0.47	177	1.60	412	1.73	3732	18.7
1977	1930	12.9	1308	4.37	115	0.38	127	1.14	368	1.55	3848	20.3
1978	2569	17.1	1312	4.38	112	0.37	140	1.26	583	2.45	4716	25.6

[a] Data from Swedish fishery statistics.

Table 2.16. Income for Fish Catch for Selected Years (Thousands of Swedish Crowns at Current Prices)[a]

Year	Herring	Cod	Flounder	Salmon	Total
1920	463	334	73	162	1122
1944	818	115	78	451	1692
1971	1038	176	138	1751	3342
1972	1645	317	143	2105	2590
1973	1160	538	187	3546	5651
1974	1576	842	142	2248	5206
1975	1177	1191	170	2568	5485
1976	1065	2015	215	4151	7727
1977	1648	2137	228	3584	7880
1978	3567	2666	280	4032	10,902

[a] Data from Swedish fishery statistics.

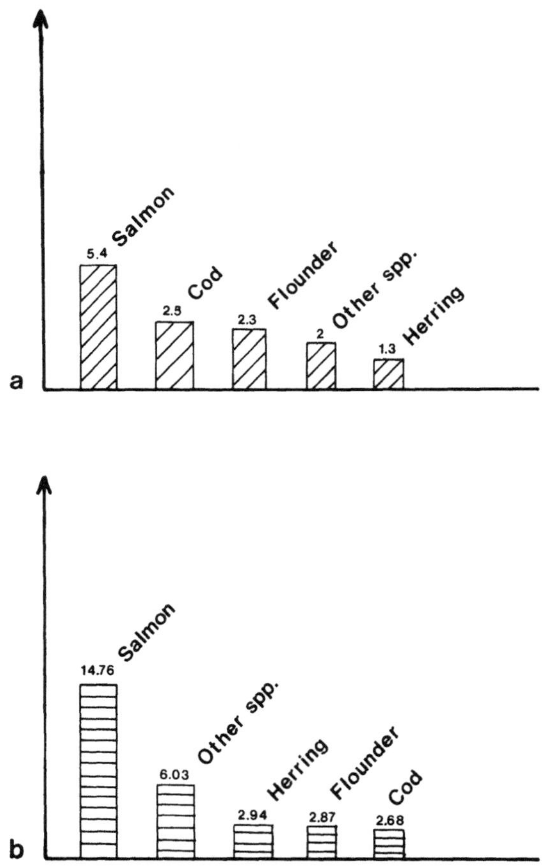

Figure 2.26. (a) Direct energy costs per energy content of catch for 1973. (b) Direct energy costs per protein energy content of catch for 1973 (see Appendix I).

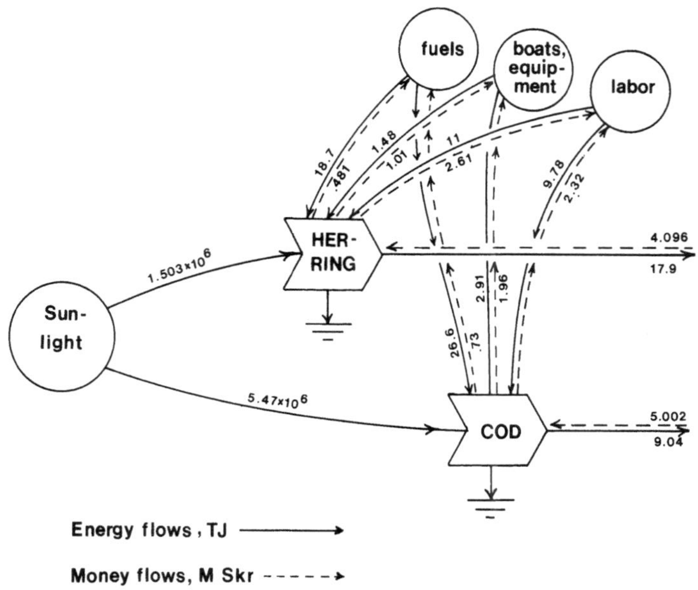

Figure 2.27. Main costs and catch for the herring and cod fisheries in 1979. The ratio of direct fuels to catch was 1.04 J/J for herring and 2.94 J/J for cod (Limburg 1983).

improved yield per worker is a function of the capital and energy intensiveness that has increased over the years. The ratio of direct energy cost to energy value of catch seems to have stabilized at a level of about 2.2 J of fossil fuel per joule of fish biomass, with the drop in 1978 attributable to large herring stocks. Estimations for the indirect energy costs of goods and materials for 1973 amounted to 28.9 TJ, which is lower than the direct energy use of 37.6 TJ (see Table 2.17). Comparison of Figures 2.26 and 2.27 for herring and cod results in estimates of direct energy cost per unit of catch— for herring, 1.3 and 1.04 J J^{-1} in 1973 and 1979; for cod, 2.5 and 2.94 J J^{-1} in 1973 and 1979. Thus, the total energy cost of 66.5 TJ resulted in an energy cost-to-catch ratio of 3.58. Any possible future expansion of fishing activity on Gotland will depend on several ecologic and economic factors, such as stocks of fish, the environmental conditions of the Baltic Sea, the competition from other countries and the mainland, the availability of energy, and the pricing structure for fish. Some of these aspects are investigated in Chapter 3 by the use of mathematic simulation models.

Forestry

Forestry has traditionally been a vital part of the economy of renewable resources on Gotland. The economically productive forest area amounts to about 1260 km^2 or 40% of the total area of the island, with pine (*Pinus silvestris*) and spruce (*Picea abies*) being the dominant species. There are other ecosystems that contain biomass in the form of trees, but presently the overwhelming majority of timber harvest takes place on the areas of merchantable coniferous

Table 2.17. Synopsis of Several Energy and Economic Parameters for the Total Gotland Fisheries

	1920	1944	1973	1974	1975	1976	1977	1978
Catch/capital (MJ/Skr)[a]	15.3	5.4	2.3	—	—	—	—	—
Catch/worker (GJ/worker)	9.6	11.6	76.2	—	—	—	—	—
Capital/worker (T Skr/worker)	0.6	2.1	33.4	—	—	—	—	—
Protein in catch in person equivalents[b]	20,660	22,100	31,060	34,125	32,140	36,060	37,060	45,060
Percentage of Gotland population whose protein demand could be met	36.9%	37.1%	57.5%	63%	59%	66%	68%	83%
Direct energy cost (TJ)	6.3	18.2	37.6	—	41.2	41.0	45.2	44.6
Indirect energy cost[c] (TJ)	—	—	28.9	—	—	—	—	—
Direct energy cost/catch	0.52	1.3	2.02	—	2.28	2.2	2.23	1.74
Total energy cost/catch[d]	—	—	3.58	—	—	—	—	—

[a] Catch expressed in energy units; capital is expressed in 1920 prices.
[b] Protein requirements assumed to be 50 g/person per day based on the age structure of Gotland's population.
[c] Indirect energy costs represent costs of goods and materials.
[d] Total energy cost = sum of direct and indirect energy costs.

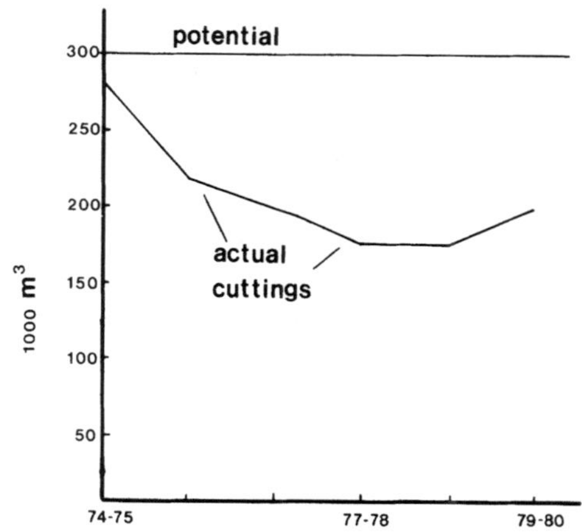

Figure 2.28. Commercial forestry cutting in recent years as comparable to maximum potential (Länsstyrelsen, 1981).

forest. Estimates of forest productivity, which are summarized in the section in this chapter on Natural Systems, amount to something on the order of the following: net production of trees = 13 PJ yr^{1}; net production of ground vegetation = 4 PJ yr^{-1}. The total gross production of forests amounted to 38 PJ yr^{-1}, which in comparison to a total incident sunlight flux of 4684 PJ yr^{-1}, represents a photosynthetic efficiency of about 0.8%. The forest system is con-

Table 2.18. Measures Characterizing the Forestry System on Gotland, 1972[a]

Parameter	Value
Economically productive forest area	1260 km^2
Sunlight incident on forest area	4815 PJ
Gross production of forest	16 tons/ha (0.3 TJ ha^{-1})
Total gross production[b]	2.03 M tons (38 PJ)
Net production of forest	7.08 tons/ha (0.13 TJ ha^{-1})
Total net production	0.89 M tons (16.5 PJ)
Total net production of trees	13 PJ
Total net production of marketable wood[c]	3.25 PJ
Total labor	234,000 man-hr
Value of sales	9.8 MSkr
Value added	7.4 MSkr
Energy value of commercial harvest	870 TJ
Fuel use by private companies	8.76 TJ
Wood cut per unit of fuel	99.3 J J^{-1}
Energy cost per value added	1.18 MJ Skr^{-1}

[a] See Appendix II.
[b] Above and below ground.
[c] Trunks represent 25% of total productivity of the tree.

nected to the human system in several ways. Energy is stored in the form of wood and edible mushrooms and in berries such as blueberries, lingonberries, and raspberries. These are gathered by local individuals and tourists. Estimates indicate that about 108,000 m^3 (870 TJ) and 40,000 m^3 (320 TJ) of wood were harvested in 1972 by private companies and individuals, respectively. Figure 2.28 presents harvests that have occurred since mid-1974, indicating that larger harvests are potentially available. Estimates were made for the fuel energy required for cutting, removing, ditching, roads, and ground preparation. This was about 8.8 TJ for private companies and 2.5 TJ for individual farmers (Table 2.18). Thus, the yield of wood in terms of energy was about 100 J J^{-1} of fuel used. Estimates were not made of indirect energy costs for goods and materials because of insufficient data. The total employment is about 150 persons, which translates into 234,000 man-hr. Of the wood harvested in 1972 by private companies, 370 TJ went to the wood industry and 500 TJ was exported, with associated economic values of 6.25 MSkr and 3.55 MSkr, respectively.

Industrial Activities

A small number of industrial sectors contribute substantially to the generation of economic value, the demand for energy, and the modification of the environment. Although resource requirements by the industrial sector are considered more formally in the section in Chapter 3 on "Input-Output Models", a more disaggregated view is presented here as well as an analysis of changes that have occurred during the period 1970–1979. It would be desirable if such data were continuously collected at the level of the firm instead of the sector in order to provide detailed information on the use of natural resources and the environmental and economic consequences that might result. Such information would also be important for allocating impacts spatially. However, data are generally not available at this level because of business confidentiality constraints; therefore, one must be content with data reported at the level of the sector to which a minimum number of three firms are assigned. For the Gotland economy, the following sectors are relevant: mining and quarries, food industries, textiles, wood industries, graphics, chemicals, stone and soil, and workshops. The most important activities, both historically and currently, include the exploitation of limestone and gravel (mining and quarries), the processing of agricultural and fishery products (food industry), the production of cement (stone and soil), the activities of the telecommunications equipment company L.M. Ericsson (workshops), and the processing of timber and production of wood products (wood industry). For 1972, it can be seen that the total energy consumption in industry represents about 35% of the total energy consumption—a significant proportion (Figure 2.17). Listed in Tables 2.19 and 2.20 are the different energy-economic characteristics of the eight sectors, with rather wide ranges in values of parameters. The dominance of energy consumption by the stone and soil sector is evident. This is not due to the absolute level of production, but rather to the energy-intensive nature of cement production. Stone and soil and food

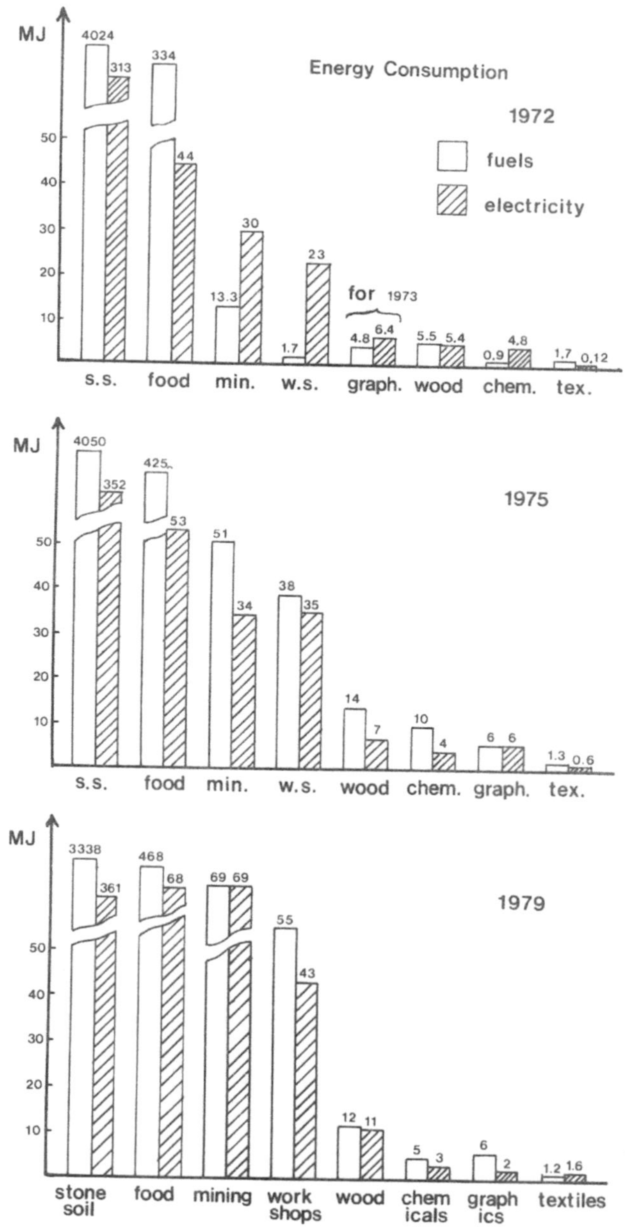

Figure 2.29. Consumption of fuels and electricity for different industry sectors in 1972, 1975, and 1979 (data obtained from industry statistics).

Table 2.19. Summary of Industry Statistics, 1972

Code	Industry Sectors	Value[a] Added (MSkr)	Value[a] of Sales (MSkr)	Wages[a] (MSkr)	Labor (t man-hr)	Fuels (TJ)	Electricity (TJ)
2	Mining and quarries	14.0	22.0	7.3	244	13.3	30.4
31	Food	55.3	278.7	29.4	1094	334.0	43.9
32	Textiles	1.2	3.3	1.1	35	1.7	0.12
33	Wood	8.3	15.9	6.9	341	5.5	5.4
342	Graphics	6.9	8.4	5.1[b]	61	4.8	6.4[b]
35	Chemicals	2.1	3.1	2.7[b]	78	0.9	4.8
36	Stone and soil	49.8	87.3	13.7	448	4024.0	313.0
38	Workshops	59.5	92.3	35.8	1717	1.7	23.3
	Totals	197.1	511.0	102.0	4018	4386	427

[a] At current prices.
[b] Value for 1973.

Table 2.20. Energy-to-Economic Measures for Industry, 1972

Code	Industry Sectors	Total Energy, E[a] (TJ oil equivalents)	E/V_A[b] (MJ/Skr)	E/V_W[b] (MJ/Skr)	E/L[b] (MJ/man-hr)
2	Mining and quarries	114.6	8.19	15.6	500
31	Food	480.3	8.69	16.3	440
32	Textiles	2.1	1.78	2.0	60
33	Wood	23.5	2.83	3.4	70
342	Graphics[c]	27.8	2.78	5.5	310
35	Chemicals[c]	17.1	9.42	6.3	140
36	Stone and soil	5067.0	101.70	370.7	11,300
38	Workshops	79.4	1.33	2.2	50
	Totals or average	5811.8	29.50	57.0	1446

[a] Total energy in oil equivalents = fuels + electricity/0.3 (see Table 2.19).
[b] V_A = value added; V_W = value of wages; L = labor in man-hr; E/V_A = energy per unit of value added; E/V_W = energy per value of wages; E/L = energy per man-hr.
[c] Data from 1973.

sectors dominate energy consumption, with mining and workshops following in that order. On the other hand, workshops and food sectors generate the most employment and wages as well as value added. Defining ratios of energy to value added and energy to labor (or wages), a comparison of energy intensiveness versus value added and total wages is possible. Thus, to generate a unit of value added, stone and soil requires 101.7 MJ, chemicals 9.4 MJ, food 8.7 MJ, and so on of direct energy consumption. With respect to the number of man-hours, we see stone and soil followed by mining, food, and other sectors (see Table 2.20). If the indirect energies associated with goods imported from

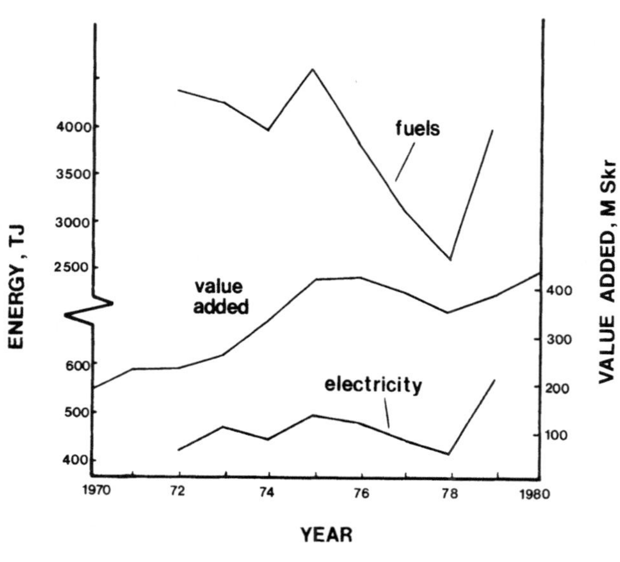

Figure 2.30. Total fuel consumption, electricity, and value added for all industry sectors combined. Value added is expressed in constant Swedish kronor at 1975 prices.

outside Gotland to the various sectors are included, then the food sector becomes more energy-intensive and the workshop sector moves to a higher position in the energy intensiveness spectrum (Jansson and Zucchetto 1978a).

The situation has changed somewhat during the period 1972–1979. In the first place, economic activity has expanded and total value added has generally risen during this period (Figure 2.30 and Table 2.21), as did total sales and wages. If the individual sectors are considered (Tables 2.22 and 2.26), then it is seen that all sectors increased in value added and sales and only the wood sector declined in total wages. The sectors that experienced the greatest percentage expansion in value added were stone and soil, chemicals, graphics, and workshops. The stone and soil expansion was due to increased production in the cement industry, while workshop expansion was due to increased output by L.M. Ericsson in many of the years between 1972 and 1980. However, employment by L.M. Ericsson fell to about 600 in 1982 from previous highs of 1200. The graphics and chemical sectors represent rather small contributions to the total economy and are not energy-intensive, so that economic expansion does not create large energy demands. On the other hand, the picture for energy demand is somewhat different, with fuel consumption declining and a generally increasing trend in the use of electricity (Table 2.24 and Figure 2.30). A rather sharp decline is observed in total energy consumption in the stone and soil sector during 1972–1979, but there generally was an increase in all the other sectors. The energy-to-value added ratios have increased in five of the eight sectors (Table 2.23 and 2.25), but they have declined in the graphics, chemical, and stone and soil sectors; for the entire industry sector, the energy to value added ratio has declined significantly over this period. Thus, industrial activity has become less energy-intensive. The significant decline in energy consumption in the cement industry has been the result of a change in the technology from a wet to a dry process, which requires significantly less energy for drying the crushed limestone. This change is also noted in Table 2.27, in which the percent of energy distribution among industrial sectors is presented; stone and soil accounts for 82% of the total energy use in 1979 compared to 90% in 1972, while all of the other sectors have risen in percent terms. On the other hand, the large contributions to wages and value added in workshops and food sectors is considerably higher than their percentages of the total energy consumption. In general, there is a reciprocal relation between energy and labor intensiveness. This is illustrated pictorially for 1972 in Figure 2.31, where the position of the different sectors is located in a graph of labor per unit of energy versus energy per unit of value added. The political importance of this fact is discussed in Chapter 4.

The Military

Although a minor part of our study, some estimates were made of the economic and energy impacts due to the presence of the military. It would certainly have

Table 2.21. Economic and Energy Measures for All Industrial Sectors Combined, 1970–1980 (Economic Measures at 1975 Prices)

Year	Value of Sales (MSkr)	Value Added (MSkr)	Value of Wages (MSkr)	Fuel per Value Added (MJ/Skr)	Electricity per Value Added (MJ/Skr)	Fuel + Electricity per Value Added (MJ/Skr)	Fuel + Electricity[a] (oil equivalent) per Value Added (MJ/Skr)
1970	265	229.6	92.8	—	—	—	—
1971	584	240.2	93.6	—	—	—	—
1972	547	242.0	93.5	18.1	1.74	19.84	23.9
1973	590	269.5	98.8	15.8	1.74	17.50	21.5
1974	769	338.6	98.7	11.6	1.30	12.90	16.0
1975	785	420.9	118.1	10.9	1.17	12.10	14.8
1976	825	425.1	—	9.0	1.12	10.10	12.7
1977	788	398.2	—	7.8	1.10	8.90	11.5
1978	745	350.5	101.1	7.4	1.16	8.60	11.2
1979	816	391.2	103.6	10.1	1.43	11.50	14.9
1980	882	441.8	122.5	—	—	—	—

[a] Fuel + electricity in oil equivalents = fuel + electricity/0.3.

Table 2.22. Value Added and Wages for Industry Sectors, 1970–80 (Millions of Swedish Crowns at 1975 Prices)

		Value Added										
Code	Industry Sector	1970	1971	1972	1973	1974	1975	1976	1977	1978	1979	1980
2	Mining and quarries	19.90	17.67	19.16	19.26	21.82	26.97	24.70	23.80	20.70	25.25	22.20
31	Food	51.90	84.80	68.20	67.09	71.05	74.30	76.60	71.10	80.83	74.18	84.72
32	Textiles	3.38	1.66	1.36	1.37	1.64	1.48	2.06	2.03	1.66	1.89	2.05
33	Wood	8.15	11.04	9.93	9.84	14.83	17.16	12.73	11.10	10.33	13.68	14.88
342	Graphics	4.71	3.60	7.54	11.77	12.04	15.28	17.35	17.23	18.28	18.72	18.95
35	Chemicals	4.42	3.53	4.31	2.14	5.31	12.29	11.93	9.57	10.71	12.73	13.97
36	Stone and soil	53.75	52.34	57.50	57.29	65.52	67.34	94.27	84.10	72.44	75.36	124.09
38	Workshops	53.36	65.54	74.04	100.17	146.38	206.02	185.50	179.20	135.56	169.35	160.92
	Totals	199.60	240.20	242	269	338.60	420.80	425	398	351	391	442

		Wages										
Code	Industry Sector	1970	1971	1972	1973	1974	1975	1976	1977	1978	1979	1980
2	Mining and quarries	6.54	6.47	6.76	6.05	6.2	6.74	6.82	6.66	7.20	7.99	7.51
31	Food	31.74	29.57	27.65	26.43	26.99	34.24	35.73	31.95	31.54	33.31	33.64
32	Textiles	1.20	1.12	1.05	1.06	1.18	0.92	1.31	1.34	1.10	1.14	1.05
33	Wood	5.54	7.05	6.80	6.38	7.02	6.94	7.30	7.51	7.34	6.87	6.63
342	Graphics	2.07	1.68	1.98	4.68	5.00	5.85	6.93	6.30	6.41	6.61	6.59
35	Chemicals	2.18	1.76	2.11	2.29	3.37	5.11	5.46	4.00	4.12	5.16	5.10
36	Stone and soil	11.25	11.67	12.51	12.86	12.92	13.98	—	—	13.89	15.83	17.28
38	Workshops	32.26	34.32	34.68	39.01	35.99	44.33	—	—	29.51	26.64	44.65(?)
	Totals	92.80	93.60	93.50	98.80	98.70	118.10	—	—	101.10	103.60	122.50

Table 2.23. Energy-to-Value Added Ratios for Industry Sectors, 1972–1979 (MJ [Heat Equivalents]/Skr at 1975 Prices)

Code	Industry Sector	1972	1973	1974	1975	1976	1977	1978	1979	Trend
2	Mining and quarries	2.28	2.34	1.87	3.15	3.40	4.16	5.65	5.47	↑
31	Food	5.54	4.96	6.84	6.43	6.27	7.24	5.95	7.23	↑
32	Textiles	1.34	0.90	—	1.28	1.21	1.23	1.45	1.48	↑
33	Wood	1.10	0.82	0.62	1.22	2.04	2.88	2.61	1.68	↑
342	Graphics	—	1.08	1.02	0.79	0.52	0.41	0.44	0.43	↓
35	Chemicals	1.32	2.76	1.43	1.14	1.42	0.84	0.65	0.63	↓
36	Stone and soil	75.40	74.60	57.97	65.37	37.90	33.15	31.09	49.08	↓
38	Workshops	0.34	0.33	0.22	0.35	0.53	0.55	0.73	0.58	↑

Table 2.24. Direct Consumption of Fuels and Electricity for Different Industry Sectors, 1972–1979 (Units are TJ)

Code	Industry Sector	1972	1973	1974	1975	1976	1977	1978	1979
				Fuels					
2	Mining and quarries	13.3	12	7.2	51	50	60	73	69
31	Food	334	289	436	425	425	457	424	468
32	Textiles	1.7	1.13	—	1.3	1.6	1.5	1.1	1.2
33	Wood	5.5	3.19	4.6	14	15	16	14	12
342	Graphics	4.8	6.30	6.2	6	6	5	6	6
35	Chemicals	0.9	1.10	1.2	10	11	6	5	5
36	Stone and soil	4024	3927	3481	4050	3247	2506	2013	3338
38	Workshops	1.7	7.3	8.1	38	57	58	52	55
	Totals	4386	4248	3944	4595	3813	3110	2588	3954

Code	Industry Sector	1972	1973	1974	1975	1976	1977	1978	1979
				Electricity					
2	Mining and quarries	30.40	33	33.5	34	34	39	44	69
31	Food	43.90	44	50	53	55	58	57	68
32	Textiles	0.12	0.1	0.1	0.6	0.9	1	1.3	1.6
33	Wood	5.40	4.2	4.6	7	11	16	13	11
342	Graphics	—	6.4	6.1	6	3	2	2	2
35	Chemicals	4.8	4.8	6.4	4	6	2	2	3
36	Stone and soil	313	349	317	352	325	282	239	361
38	Workshops	23.3	26	24.6	35	42	41	47	43
	Totals	421	468	442	492	477	441	405	559

Table 2.25. Energy-to-Value Added Ratios for Industry Sectors, 1972–1979 (MJ [Oil Equivalents]/Skr at 1975 Prices)

Code	Industry Sector	1972	1973	1974	1975	1976	1977	1978	1979	Trend
2	Mining and quarries	5.80	6.33	5.45	6.09	6.61	7.98	10.60	11.84	↑
31	Food	9.25	6.50	8.48	8.10	7.94	9.15	7.60	9.36	—
32	Textiles	1.54	1.07	—	2.23	2.23	2.38	3.27	3.46	↑
33	Wood	2.37	1.82	1.34	2.18	4.06	6.25	5.55	3.56	↑
342	Graphics	—	2.35	2.20	1.70	0.92	0.68	0.69	0.68	↓
35	Chemicals	3.92	8.00	3.99	1.90	2.60	1.32	1.09	1.18	↓
36	Stone and soil	88.13	88.85	69.26	77.57	45.96	40.98	38.80	60.26	↓
38	Workshops	1.07	0.94	0.62	0.75	1.06	1.09	1.54	1.17	↑

Table 2.26. Percent Change in Economic and Energy Variables for Industry Sectors, 1972, 1978, and 1979[a]

Code	Industry Sector	Fuel Use	Electricity	Value Added	Wages	Sales	Cost of Raw Materials	Energy per[b] Value Added
2	Mining and quarries	+434	+86	+24	+14.6	+30	+78	+44
31	Food	+34	+42	+17	+21	+8.9	+7.5	+19
32	Textiles	−32	+1108	+45	+4.3	+83	+177	+9
33	Wood	+136	+122	+44	−0.7	+91	+183	+95
342	Graphics	+25	−69	+151	+233	+178	−60	−60[c]
35	Chemicals	+456	−48	+210	+143	+199	+229	−52
36	Stone and soil	−34	−4	+247	+32	+29	+29	−47
38	Workshops	+3047	+93	+123	+2.8	+119	+152	+93
	All industry	−25	+14	+53	+9.5	—	—	−42

[a] Average for 1978 and 1979 in comparison to 1972. Economic measures in 1975 prices (see Table 2.22).
[b] From Table 2.25, with energy expressed in oil equivalents.
[c] Change from 1973.

Table 2.27. Percent of Total Industry Attributable to Different Industry Sectors for Energy Consumption, Wages, and Value Added, 1972, 1975, and 1979

Code	Industry Sector	1972			1975			1979		
		Energy	Wages	Value Added	Energy	Wages	Value Added	Energy	Wages	Value Added
2	Mining and quarries	0.90	7.20	7.90	1.67	5.70	6.40	3.06	7.7	6.50
31	Food	7.86	29.60	28.20	9.40	29	17.70	11.90	32.2	19
32	Textiles	0.04	1.12	0.56	0.04	0.78	0.35	0.06	1.1	0.50
33	Wood	0.23	7.30	4.10	0.41	5.90	4.10	0.50	6.6	3.50
342	Graphics	—	2.10	3.10	0.24	4.95	3.60	0.18	6.4	4.80
35	Chemicals	0.12	2.26	1.80	0.28	4.30	2.90	0.18	5.0	3.26
36	Stone and soil	90.20	13.40	23.80	86.50	11.80	16	82	15.3	19.30
38	Workshops	0.52	37	41.40	1.44	37.50	49	2.18	25.7	43.30

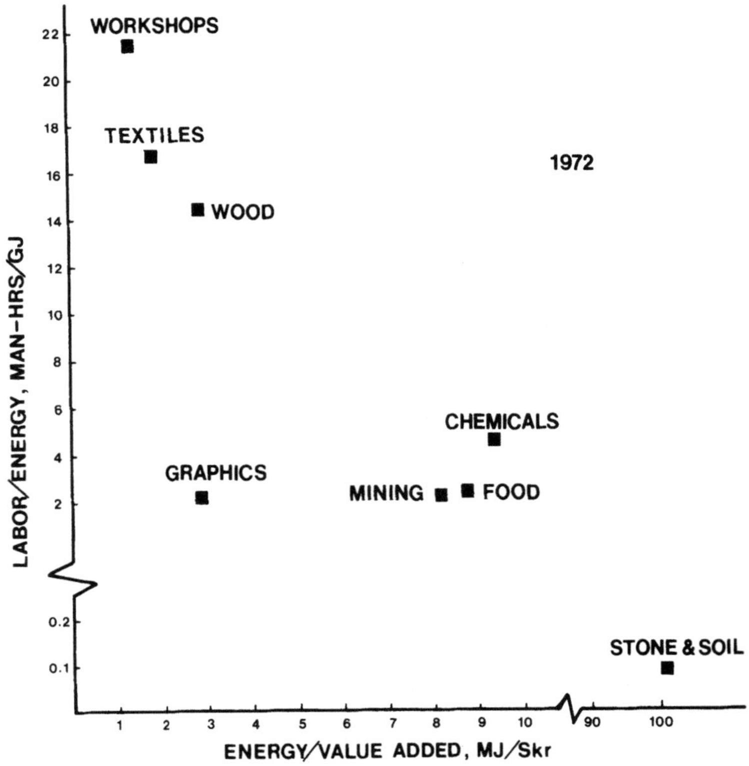

Figure 2.31. The generally inverse relationship between labor per unit of energy consumed and energy consumed per unit of value added generated for 1972.

Table 2.28. Summary of Parameters Characterizing Military Activities on Gotland, 1972

Parameter	Value
Number of servicemen	1750
Full-time employees	1200
Wages generated	45 MSkr
Municipal taxes generated	11 MSkr
Total military consumption	179 MSkr
Electrical consumption	140,000 kWh or 0.5 TJ
Diesel fuel consumed	2100 m^3 or 74.8 TJ
Fuel oil no. 1 consumed	4000 m^3 or 142 TJ
Gasoline consumed	1400 m^3 or 44 TJ
Food consumed	739 tons (7.5 MSkr or 11 TJ)

been of interest to assess the environmental impacts of military exercises and maneuvers, but this would require open access to military information. About 1750 servicemen per year are stationed on the island, 500 of whom are native Gotlanders. Military activities generate about 1200 full-time jobs, which yielded 45 MSkr in wages and 11 MSkr in municipal taxes in 1972. The total military consumption of 179 MSkr in 1972 was a substantial part of the GRP of 1158 MSkr. A more extensive analysis of the military system might be an interesting example of order-disorder processes, with the energy of explosives and weapons resulting in direct destruction to the environment, while the influx of subsidies from the central government for the military supports the level of living for those working for the military. A summary of the flows of energy, goods, and money associated with the military on Gotland in 1972 is presented in Table 2.28.

Tourist Activity

Tourism constitutes a significant aspect of the region and influences the overall character and resource use of the island during the summer season. The number of tourists has steadily increased in the post-World War II period, and tourist visits were estimated to be 300,000 in 1982 compared to about 200,000 in 1972, which is an increase of about 50% (Figure 2.32). Tourists primarily come from mainland Sweden in June, July, and August for camping, swimming, and other outdoor activities. Access to the island has been made easier with the operation of a large car ferry in 1981 in addition to the three car ferries that were already operating. Future levels of tourism will also depend on factors such as the state of the Swedish and world economies, the wealth of the average Swedish citizen, the price of fuel, and the exchange rate of the Swedish crown, all of which will determine whether Swedes can afford tourism in other parts of the world. If the Swedish economy declines and the cost of international air transportation increases, then one can expect that Gotland will become even more attractive to Swedes as a vacation area. Finally, there is a growing interest to make Gotland

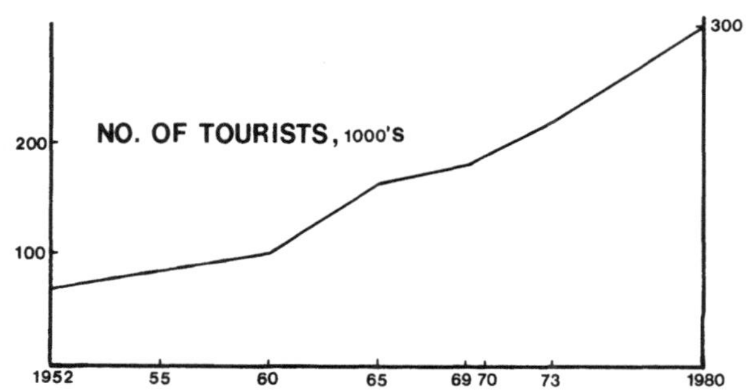

Figure 2.32. Tourist visits, 1952–1980 (data obtained from Länsstyrelsen 1975, and recent data from the Planning Department on Gotland).

more attractive to visitors from other Nordic countries and the European continent. The number of visitors from Denmark, Finland, and West Germany has steadily increased due to increasing numbers of ferry routes and available accommodations on the island.

The influx of 300,000 people over a 2–3-month period contributes economic benefits, with tourist expenditures stimulating activities tied to providing goods and services for visitors. However, wherever one has large groups of people, the problems associated with congestion and stresses arise. This is very much the case in and around Visby due to its being the principal point of entry and departure by sea and air. The significant expansion of the harbor in Visby to supply enough landing space for the new ferry has significantly changed the local area around the harbor. Visby itself is one of the most attractive Swedish tourist sites, with its striking architecture, medieval ambience, and agglomeration of activities. In summer, the old city with its narrow cobblestoned streets, is mostly crowded with tourists. There are also large campsites for mobile homes along the coast just outside the downtown area, that are jammed in the summer. Scattered around the island are other local points for camping or tourist activity, mostly along the coasts, which in some cases have led to the establishment of communal housing developments for vacationers. The vast majority of summer houses, however, are privately owned by inhabitants of the mainland.

Attention has thus far been focused on the direct locational impacts. However, a more far-reaching potential impact is derived from an issue that has already been mentioned; namely, the availability of water. The timing of tourist activity is important in this respect, because summer is a relatively dry period with high rates of evapotranspiration. A significantly increased demand for water may lead to declining groundwater levels and salt water intrusion in some areas, especially along the coast. This may eventually result in pressure to expand water supplies by using surface water from a lake situated far up north, as well as by building dams to increase surface storage capacities. These efforts, if they come to pass, will consequently require investments of resources and energy for meeting demand during the tourist season; thus, a trade off between benefits and costs of various levels of tourist activity arises. Estimates for 1972 indicated that income from direct tourist payments for services was about 40 million Skr, constituting about 6% of the total income for exports (including transfer payments) of 658 million Skr (note that the GRP in 1972 was about 1160 million Skr.). For 1975, it was estimated that the total output from the tourist sector amounted to 80 MSkr (at 1975 prices); it contributed a value added of 71.4 MSkr, which is 4.6% of the GRP. This resulted in the employment of 2270 people and the consumption of 344 TJ of fuels and 15 TJ of electrical energy.

Transportation

The links between various components in a society are provided by systems of transportation and communication for the interchange of energy, goods, people, and information. For the most part, regional studies are concerned with transportation activities, recognizing that transport connections within—as well as to and from—a region are important for economic and cultural activity. A complete study of the transportation system on Gotland was not conducted during our project, but several energy and economic aspects of this activity have been assessed for the purposes of energy planning and inclusion in the input-output models, which are presented later. Transportation costs to and from Gotland are ameliorated by a system of subsidies from the central government in an attempt to make local industries competitive with those on the mainland. Access to the island by tourists and their cars is also of prime importance for the tourist economy. This has been made possible by an expansion of the summer ferry service, with a significant increase of capacity after the addition of a large ferry in 1981. Goods arrive on these ferries aboard truck trailers and are distributed on Gotland from a central terminal in Visby. The major transport into Gotland consists of fossil fuels, fertilizer, feed and food products, as well as vehicles and tourists. Major economic exports consist of limestone, cement, wood, industrial goods, food, and other assorted goods. Four large transport companies handle the bulk of these transports; one specializes in consumer goods and provisions, one in heavy transports, one deals in consumer goods, and (finally) one in the intraregional flows of bulk and piecemeal commodities as well as lumber. Bulk commodity flows on the island are mostly by truck since the railway system was discontinued in 1960.

There is an extensive system of county and municipal roads with a total length of 2200 km throughout the island. Commuting between rural areas and Visby for work and social visits makes up a significant part of the automobile traffic. Approximate values for the number of local vehicles in 1972 and 1979, as well as the consumption of gasoline and diesel fuel in transportation (1972), are summarized in Tables 2.29 and 2.30. Fuel consumed in transportation

Table 2.29. Number of Vehicles (1972 and 1979) and Consumption of Diesel Fuel (1972)[a]

Category	No. of Vehicles 1972	No. of Vehicles 1979	Total Energy Consumption 1972 (TJ)
Private cars	372	—	33.1
Taxis[b]	60	—	5.3
Trucks	525	736	—
Åkericentralen	135	—	94
Dairies and Farmek	16	—	20.6
Municipalities	7	—	7.8
Others	367	—	92.6
Buses	54	89	27
SJ Buses	18	—	9.6
Total			290

[a] Does not include military or agriculture (see Jansson and Zucchetto 1978a).
[b] Number included in total.

Table 2.30. Number of Vehicles (1972 and 1979) and Consumption of Gasoline (1972)[a]

Category	No. of Vehicles 1972	No. of Vehicles 1979	Total Energy Consumption 1972 (TJ)
Private Cars	15,345	20,236	581
Commuting	3300[b]	—	63
Tourists	37,000	—	88
Truck Transport	460	660	38
Municipality	6	—	3.1
Military	—	—	44
Buses	35	15	16
Motorcycles	400	—	—
Motorbikes and machinery	5000	—	9.4
Other	—	—	163.3
Total			1006

[a] Data for 1972 from Jansson and Zucchetto 1978a.
[b] Included in private car totals.

amounted to about 73% of the total liquid fuels imported in 1972. Commuting between Gotland and the Swedish mainland is mainly by air. Air travel to and from Visby has increased from 200,000 passengers in 1972 to 275,000 in 1981–1982.

Urban Sector

The flows of materials and energy within the urban sector were not studied in any detail in reference to the limited scope of our project; in principle, however, this would have been an interesting subject of study in itself (Zucchetto 1975a, 1975b). In 1972, the urban population (defined as the total population minus the farm population) was 42,600; this population is presently about 43,500. Data on energy consumed in this sector for 1972 are contained in Table 2.31 by type

Table 2.31. Energy Consumption in the Urban-Suburban Sector by Type, 1972[a]

Energy	Quantity		Energy Value		Percent of Imports[b]
	m^3	m^3/Person	TJ	GJ/Person	
Gasoline	24,144	0.567	757.2	17.77	74
Diesel	11,344	0.266	406.2	9.54	56
Oil No. 1	19,944	0.468	710	16.67	20
Oil No. 3	14,832	0.348	577	13.54	94
Oil No. 4	1647	0.039	64	1.50	18
Totals	71,911	1.688	2514.4	59.02	41
Electricity	163.3 GWh	3.83 MWh/person	588	13.80	108

[a] Data from Jansson and Zucchetto (1978a).
[b] Percentages calculated based on volume (m^3) of each fuel imported.

Table 2.32. Electrical Energy Consumption in Several Urban Activities for Selected Years (MWh)

Activity	1972	1975	1978
Building and work activities	2134	3266.9	4621
Waterworks	2792.4	3273.7	3406
Sewage treatment	1583.3	2442.4	2605
Wholesale	885.2	1139.8	1781
Retail	14,803.5	14,805.2	19,760
Banks	452.5	458.2	975
Street lighting	4827	5252.9	6063

and also by percentage of total imports to the island. A majority of gasoline, diesel fuel and oil no. 3 was consumed in the urban system and more electricity was consumed than was imported via the cable in 1972. The generally increasing trend of electricity consumption (Figures 2.13 and 2.14) is notably manifest in activities connected to the urban-suburban areas (Table 2.32).

There has been a general urbanization trend during the post-World War II period, with the greatest absolute change taking place in Visby and its immediate surroundings (Table 2.33). The part of Visby within the old walls, or the old city, has remained relatively untouched. However, expansion has proceeded outside the wall with the introduction of new apartments, shops, supermarkets, and road nets (Figure 2.6). In spite of this, congestion is problematic during the tourist season, when the effective population is greatly enlarged.

The Study Site of Lummelunda

During the last phase of the project, attention was focused on a small subarea of Gotland; namely, the 60-km^2 drainage basin around Lummelunda, north of Visby. The major part of this drainage basin consists of the former Martebo mire, which is now drained and cultivated. Arable land makes up about 53% of the basin compared to 26% for the entire island. The site was chosen to make more demonstratively evident techniques that were being used in the project. These included a 1-year monthly sampling of hydrologic data for the Lummelunda creek to validate a hydrology model, which was a modification of the model developed earlier in the Gotland project (see the section in Chapter 3 on "Model 6"). It was used to assess water quality in the watershed due to land-use change and intensified agricultural activity (Nilsson 1982). A field investigation of vegetation types was undertaken to ground truth and validate aerial photographs of the area to assess land-use change that had occurred in the drainage basin between 1958 and 1980 (Nordberg 1983, Alm and Nordberg 1983). It was initially anticipated that satellite remote sensing techniques would be used to give an environmental inventory, at least with respect to land-use and vegetation for the entire island. The Lummelunda area, together with a

Table 2.33. Population and Growth in Area of Selected Towns on Gotland

Town	Population in 1972	Area in 1961 (ha)	Area in 1972 (ha)	Change in Area		Rate of Change
				(ha)	(%)	(ha yr^{-1})
Visby	19,409	234	394	160	68	14.5
Slite	1747	42	58.8	16.8	40	1.5
Hemse	1221	37.8	58	20.2	53	1.8
Klintehamn	1212	22.5	54	31.5	140	2.9
Romakloster[a]	1036	14	42.8	28.8	205	2.6
Fårösund	954	27.5	46.5	19	69	1.7
Vibble[a]	573	8.8	33	24.2	275	2.2
Lärbro	556	—	—	—	—	—
Havdhem	320	—	—	—	—	—

[a] Romakloster is located right next to Roma; Vibble is located 5 km south of Visby.

number of other representative areas from different parts of the island, could then be used to calibrate the less detailed satellite pictures. However, the time and costs needed to develop a computerized interpretation system for constructing land-use maps for the entire island turned out to be much greater than anticipated. In general, the results for the Lummelunda area showed a decrease in wooded area due to clear-cutting and an ongoing reclamation of wetlands. Improved drainage and irrigation has increased productivity of existing arable land; however, at the same time, it has led to a decline in the diversity of the rural landscape.

Plate 2. The old town of Visby is the dominant urban center, a medieval walled city with an administration center and main harbor (Photo, B. O. Jansson).

3. Modeling Approaches and Results

Introduction

Systems models have been a major methodologic part of many areas of study in science and engineering; they have certainly been an important focal point for our entire study. Over the period during which the Gotland study has been active, several quantitatively oriented models have been formulated. These models were modified and changed as data became available, different concerns gained precedence, or new ideas emerged. Although one of the aims of the Gotland project was to describe the island in some holistic systems formulation, this did not imply that our intention was to generate a very large and complex model that could incorporate all aspects of the entire system. It was felt that such an approach would rapidly become intractable, inflexible, and self-defeating. However, in the first phase of the project, an attempt was made to evaluate the major flows of energy, materials, and money within and between the natural and human systems to obtain an overall quantitative perspective for the entire island (Jansson and Zucchetto 1978a). At the time, the analysis included a mathematic simulation model incorporating hydrology and water quality considerations. Since then, the integration of different modeling efforts has become a main strategy of the project (Figure 3.1). Although it was initially anticipated that differential equation models would be used for all modeling efforts, the fact that many storages, such as capital, change relatively slowly in the economic system and that only certain types of economic data were

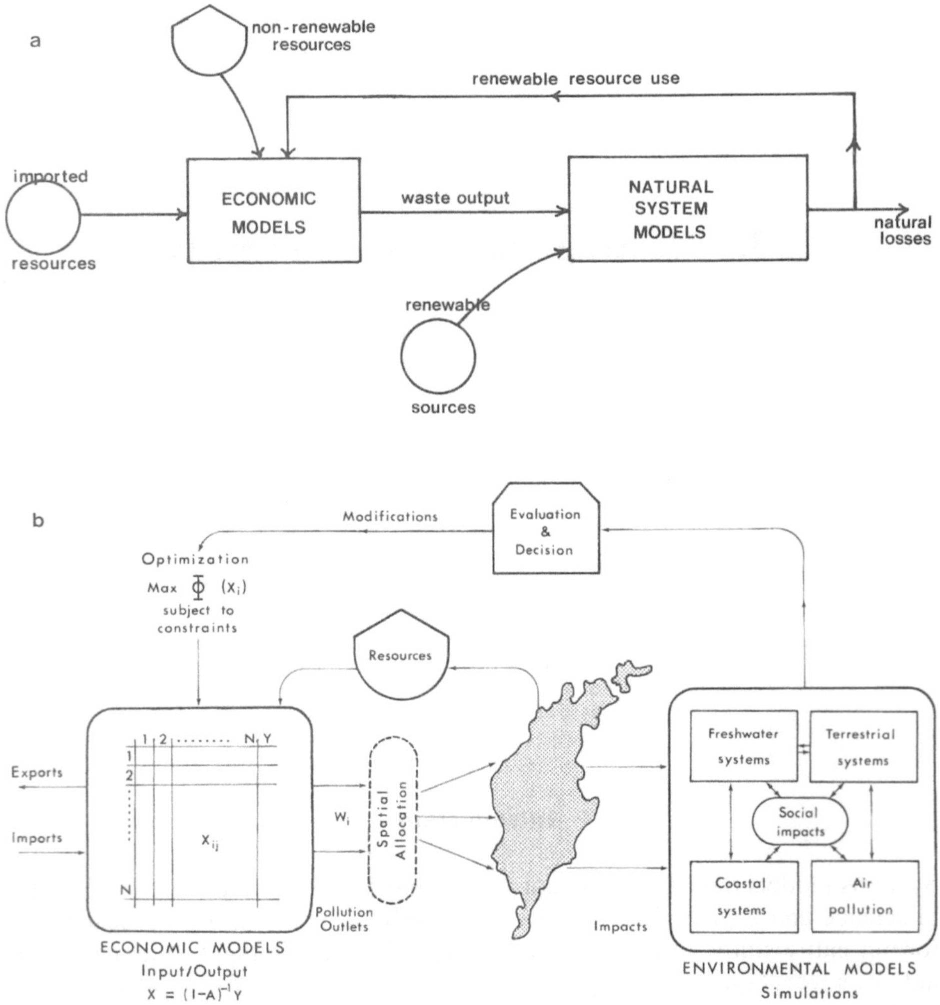

Figure 3.1. Coupled set of economic-ecologic models for total system analysis. An economic model or prescription of activity would allow estimates of resource demands and waste production. These estimates would be used to assess and evaluate resource and ecologic impacts with natural system models. Evaluated impacts would result in a reassessment of economic activity. (a) General organization. (b) Gotland project organization.

reasonably accessible prompted us to adopt an input-output formulation for quantitative representation of the economic system; these are discussed later in this chapter in more detail. Such models could be used (at least over short time periods) to estimate outputs of each sector based on anticipated final demand, to assess the impact on economic output for shortages of inputs such as energy, and to generate economic and resource multipliers. They could also be incorporated into optimization formulations, in which some objective function is being maximized or minimized. In general, in the economic model, the regional system interacts with the rest of the world by exchanging imports and exports and consumes storages of indigenous resources. Wastes and impacts are produced that can be used as inputs to various ecologic-environmental models to assess their influence on the environmental systems. When this information is evaluated, modifications can be considered for the economic system that would result in an acceptable level of predicted environmental impact and hopefully would stimulate action against environmental degradation.

This interaction between economic and environmental concerns was representative of the Gotland project organization with respect to models and analysis, but it could also serve as a paradigm within an actual planning and decision framework. A project, which is composed of people engaged in interaction, is an evolving organic undertaking. A mechanistic view of the modeling organization (Figure 3.1) perhaps would entail the notion that the total system is represented by a number of interacting, mathematically formulated submodels; however, it should be emphasized that each model is a combination of a human element (the individual scientist) and some quantitative representation of the system. The researcher formulates a model to deal with complexity as a research and intellectual aid; however, there is constant interaction between the investigator's experience, empiric knowledge, and the results of the model. A synthesis emerges that is passed on to other investigators for consideration in their respective models. For a long-term project (such as the Gotland study), investigators change, models are reformulated, and new issues and data arise. The total modeling picture becomes an extremely dynamic and evolutionary enterprise. However, in some fashion, the project leaders must try to maintain some reasonably coherent overall direction for the project. As far as the Gotland project is concerned, several analytic models were formulated involving economic and energy concerns. Analytically oriented environmental models were formulated with regard to the effects of changing land use on hydrology and contamination of water by nitrates, and also to the ecology of the coastal system. Atmospheric dispersion models were used to evaluate the impact of emissions from the cement factory in Slite. The results indicated no significant pollutant concentration of sulfur dioxides, and they are not included in the present work. In addition, several quantitative models for terrestrial ecosystems, such as forests or agriculture, were formulated but not analytically tested (see Chapter 2), even though they were included in the long-range plans of our project. Due to the lack of scientific expertise, specific social impact models were not included. For those investigators not familiar with the modeling methods used, introductions are included to input-output methods, linear optimization, and systems simulation for energy and ecology. All of these methods were used to one extent or another in our analyses.

In the subsequent sections of this chapter, the models formulated during the Gotland project are succinctly summarized and described; further details can be found in the "References" section for some of the models. Models are also presented which were formulated early in the project, even though they may have been eventually superceded by alternative formulations containing better

Table 3.1. Input-Output Transactions Table

	Purchasing Sectors	Local Final Demand				
To / From	1 ··· j ··· n	House-holds	Private Investment	Govern-ment	Exports	Total Gross Output
1	X_{11} ··· X_{1j} ··· X_{1n}	C_1	I_1	G_1	E_1	X_1
·	· · ·	·	·	·	·	·
·	· · ·	·	·	·	·	·
·	· · ·	·	·	·	·	·
i	X_{i1} ··· X_{ij} ··· X_{in}	C_i	I_i	G_i	E_i	X_i
·	· · ·	·	·	·	·	·
·	· · ·	·	·	·	·	·
·	· · ·	·	·	·	·	·
n	X_{n1} X_{nj} X_{nn}	C_n	I_n	G_n	E_n	X_n
Labor	L_1 ··· L_j ··· L_n	L_c	L_I	L_G	L_E	L
Other value added	V_1 ··· V_j ··· V_n	V_c	V_I	V_G	V_E	V
Imports	M_1 ··· M_j ··· M_n	M_c	M_I	M_G	—	M
Total gross outlay	X_1 ··· X_j ··· X_n	C	I	G	E	X

data. For example, a more complex input-output model replaced an earlier version. Although the models may seem to be preliminary in that data uncertainties do not allow the generation of precise predictions, both the issues addressed and formulations arrived at served as a means for directing further research and for highlighting certain types of model behavior. We also think they can serve as guides for issues that are of importance in other regions. The more comprehensive models, with regards to the regional system, are presented first and the smaller subsector models, such as fisheries and hydrology, are presented towards the end.

Input-Output Models

A rather straightforward class of models that have been used extensively in economic and energy studies are input-output models (Leontieff 1966, Isard 1972, 1975, Richardson 1972). These models represent a useful accounting scheme for the systematic analysis of interactive effects among a group of sectors in an economy that are due to either some change in demand for the output of these sectors or to the inputs to these sectors. These models have also served as a useful calculation scheme for determining the total embodied energy cost of various goods, for predicting the indirect impacts on industry activity due to future demand in a given economy, and for estimating indirect pollutant generation in various sectors of an economy. They are not phenomena or expressions of fundamental laws or behavior except for conservation of money, energy, or matter. The most critical assumption is that sectoral input is linearly related to output. Nevertheless, they are useful approximations for determining systemic effects of a material nature in the short-run. An input-output model of an economy usually is constructed by assuming that the economy can be divided into n sectors, which have transfers among them. Each sector produces a certain output that can go to other sectors, to local final demand (households, investment, and government), or to export. Referring to Table 3.1, each sector can have a conservation equation of the following form written for it by summing across a row, *i*:

$$X_{i1} + X_{i2} + \ldots + X_{ij} + \ldots X_{in} + C_i + I_i + G_i + E_i = X_i$$

$$i = 1,2, \ldots n$$

where X_{ij} is the transfer from sector i to sector j.

$C_i + I_i + G_i$ = sum of household, private investment, and government purchases from sector i.

E_i = exports from sector i.

X_i = total output from sector i.

For each equation, the same units for X_{ij}, C_i, I_i, G_i, and E_i must be used; in many instances, money flow is the numeraire. The above equation can be written in the somewhat simpler form of:

$$\sum_{j=1}^{n} X_{ij} + Y_i = X_i \qquad i = 1,2,\ldots n$$

where $Y_i = C_i + I_i + G_i + E_i$ = final demand.

Now, the crucial assumption made is that the transfer from a producing sector to a purchasing sector is a linear function of the output of the purchasing sector, so that:

$$X_{ij} = a_{ij}X_j$$

or

$$a_{ij} = X_{ij}/X_j$$

where a_{ij} = the direct technical input coefficient. Substituting this in the conservation equation yields:

$$\sum_{j=1}^{n} a_{ij} X_j + Y_i = X_i \qquad i = 1,2,\ldots n$$

These n equations can be written in the following matrix notation:

$$\underline{\underline{\mathbf{A}}}\mathbf{X} + \mathbf{Y} = X$$

where $\underline{\underline{\mathbf{A}}}$ = nxn matrix of technical coefficients, a_{ij}.

$\mathbf{X}$ = column vector of sector total outputs.

$\mathbf{Y}$ = column vector of final demand for output from each sector.

If we let $\underline{\underline{I}}$ = the nxn identity matrix, then this equation can be solved for the required outputs, **X**, of the different sectors:

$$\mathbf{X} = (\underline{\underline{I}} - \underline{\underline{A}})^{-1}\mathbf{Y}$$

allowing calculation of the required sectoral outputs for some level of final demand. $(\underline{\underline{I}} - \underline{\underline{A}})^{-1}$ is usually referred to as the Leontieff inverse. The model is made operational by having a set of flow data for a given year, and by calculating the matrix of technical coefficients, $\underline{\underline{A}}$; then, by assuming that if the relationships between sectors stay the same, a new final demand will allow calculation of sector outputs. Thus, if exports or household consumption is expected to increase, this will result in a new level of outputs, X_i. This change in sector activity will have labor, income, land-use, and environmental effects. If one knows the relationship of these parameters of interest to sector outputs, X_i, estimates of changing regional conditions can be made. These models can be updated in the sense of revising technical coefficients as economic behavior and technology change. Environmental effects could be included by defining a pollutant generation matrix, which would prescribe the amount of pollutant "*i*" generated per unit of output from sector j (Zucchetto et al. 1980b):

$$\underline{\underline{P}} = [P_{ij}]$$

where P_{ij} = amount of pollutant *i* per unit of output from sector "*j*."
$\underline{\underline{P}}$ = *mxn* matrix, *m* pollutants, and *n* sectors.

The total amount of wastes generated would be given by:

$$\mathbf{W} = \underline{\underline{P}}\mathbf{X}$$

where $W = m \times 1$ vector of the total amounts of each pollutant generated. This type of input-output approach could also be made spatial by considering "*s*" regions, each of which has n sectors; these sectors interact not only within a region, but between regions (Richardson 1972). This would lead to an interregional input-output model in an attempt to capture the effects of changes in one region on itself as well as on other regions. This prescription outlined above is then one of: (1) estimating changes in final demand of outputs from given sectors based on such things as export demand, increasing population, increasing standard of living, and changes in investment or planning goals; (2) translating this change in final demand to systematic impacts throughout the economy in terms of changes in sector outputs; and (3) relating these predicted changes to social, geographic and environmental aspects of interest. This is a demand-controlled model; even if there is great confidence in the evaluation of the model parameters for a given economy, it is still necessary to predict final demand.

In many circumstances, however, interest might focus on the limitations to the level of output for a given economic system: this, in a sense, has been a major theme of the Gotland study in terms of major limiting factors (such as energy) on the level of activity in the natural and economic systems. Considering the transactions part of Table 3.1, a summation down each column results in the following set of equations (Ghosh 1958):

$$X_{1j} + \ldots X_{ij} + \ldots X_{nj} + \Delta V_j + M_j = X_j \quad j = 1,2,\ldots n$$

where X_{ij} = transfer from sector i to sector j.
ΔV_j = value added to sector j = labor plus other value added.
M_j = imports to sector j.
X_j = total outlay for sector j = total output.

or

$$\sum_{i=1}^{n} X_{ij} + \Delta V_j + M_j = X_j \quad j = 1,2,\ldots n$$

Now, if it is assumed that intersectoral flows are proportional to the output of the producing sectors, then:

$$X_{ij} = \bar{a}_{ij} X_i$$

so that:

$$\sum_{i=1}^{n} \bar{a}_{ij} X_i + \Delta V_j + M_j = X_j \quad j = 1,2,\ldots n$$

$$\text{or} \quad \mathbf{X}' \underline{\underline{\bar{\mathrm{A}}}} + \Delta\mathbf{V}' + \mathbf{M}' = \mathbf{X}'$$

$$\text{where } \underline{\underline{\bar{\mathrm{A}}}} = \begin{vmatrix} \bar{a}_{11} & \bar{a}_{12} & \ldots & \bar{a}_{1n} \\ \bar{a}_{21} & \bar{a}_{22} & \ldots & \bar{a}_{2n} \\ \cdot & \cdot & & \\ \cdot & \cdot & & \\ \cdot & \cdot & & \\ \bar{a}_{n1} & \bar{a}_{n2} & \ldots & \bar{a}_{nn} \end{vmatrix}$$

$\Delta\mathbf{V}'$ = row vector of value added to sectors.
$\mathbf{M}'$ = row vector of imports to sectors.
$\mathbf{X}'$ = row vector of outputs from each sector.

Thus,

$$\mathbf{X}' (\underline{\mathrm{I}} - \underline{\underline{\mathrm{A}}}) = \Delta\mathbf{V}' + \mathbf{M}'$$
$$\text{and } \mathbf{X}' = (\Delta\mathbf{V}' + \mathbf{M}') (\underline{\mathrm{I}} - \underline{\underline{\bar{\mathrm{A}}}})^{-1}$$

This formulation leads to the determination of the output levels from each sector as a function of inputs and value added; if there are constraints on these inputs, then the output level of each sector will be constrained. This type of formulation is of interest when addressing the effects of resource shortages.

Input-output analysis can also be used to calculate direct and indirect resource requirements per unit of output from a given sector. Bullard and Herendeen (1977) used this approach to calculate embodied energy coefficients for the U.S. economy. First, define an embodied energy coefficient, e_j, which represents the direct and indirect energy cost per unit of output from a given sector, j. Then an energy balance for each sector yields:

$$\sum_{i=1}^{n} e_j X_{ij} + E_j = e_j \qquad j = 1,2,\ldots n$$

$$E_j = e_j - \sum_{i=1}^{n} e_j X_{ij} \qquad j = 1,2,\ldots n$$

In matrix form:

$$\mathbf{E}^1 = \mathbf{e}^1 (\underline{\underline{\hat{X}}} - \underline{\underline{X}})$$

$$\mathbf{e}^1 = \mathbf{E}^1 (\underline{\underline{\hat{X}}} - \underline{\underline{X}})^{-1}$$

where $\mathbf{E}^1$ = row vector of external energy contributions to each sector (energy units).

$\mathbf{e}^1$ = row vector of embodied energy coefficients for each sector, i.e., direct and indirect energy cost per unit of output from each sector.

$\underline{\underline{\hat{X}}}$ = diagonal matrix of outputs from each sector.

$\underline{\underline{X}}$ = matrix of intersectoral flows, where each element X_{ij} is flow from sector i to sector j.

Dynamic Systems Models

Fundamental descriptions of the dynamics and behavior of the observable world—in such fields as biology, ecology, economics, engineering, and physics—are couched in the form of differential equations, which describe the dependence of variables on space and time derived from fundamental laws that are subject to empiric verification. In this context, differential equations become the language of description that generates precise and measurable results. The presentation and understanding of models of ecosystems and economic systems is facilitated for some investigators by using a symbolic language to represent the system interactions being mathematically described by the differential equations. Many of the models discussed later in this volume use the symbolic language of "energese" developed by H.T. Odum (1971, 1972) for the representation of ecosystems. The basic notion is that systems can be described in terms of storages or state variables, flows between these state variables, and flows to and from the environment of the system. A description of the basic symbols employed is presented in Figure 3.2 and a more precise discussion of these symbols follows.

Source

The source is a flow of energy, matter, or information from outside the boundaries of the system into the system. It can be thought of as a causal force,

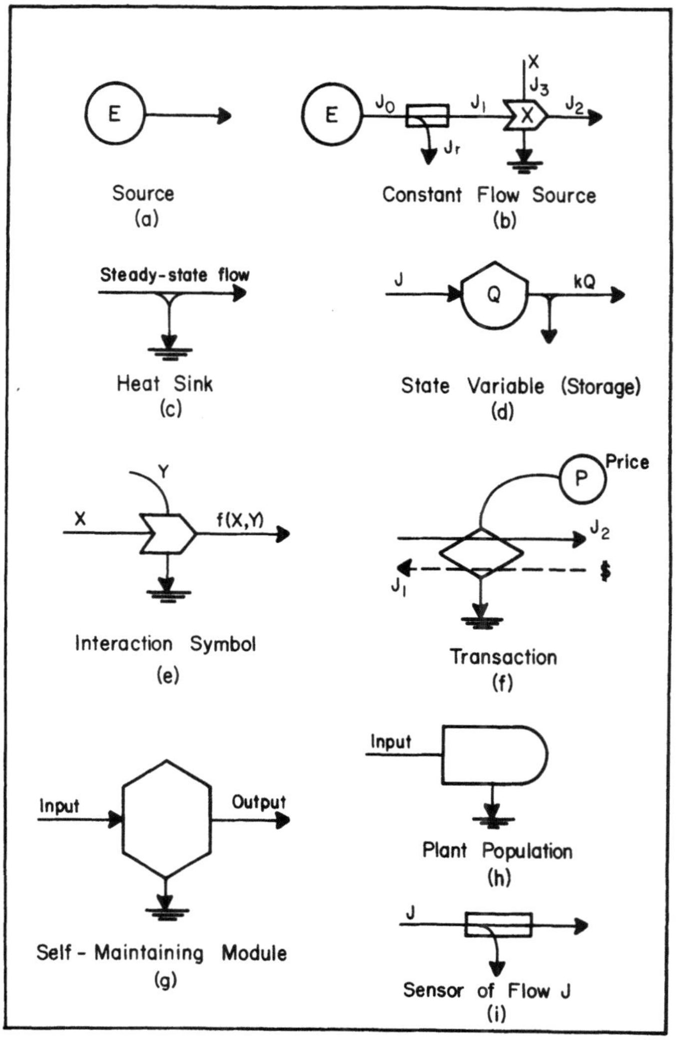

Figure 3.2. The symbols of the energy circuit language. (a) Outside source of energy supply to the system controlled from outside; a forcing function (E). (b) Constant flow source from outside; $J_r = J_0 - k_0 J_r X$, $J_2 = k_1 J_r X$, $J_2 = k_1 J_0 X/(1 + k_0 X)$. (c) A pathway whole flow is proportional to the quantity in the storage or source upstream ($J = k_1 E$). The heat sink represents the energy losses associated with friction and backforces along pathways of energy flow. (d) Storage of some quantity in the system. The rate of change equals inflows minus outflows ($\dot{Q} = J - kQ$). (e) Interaction of two flows to produce an outflow that is some function of these flows—usually a multiplicative output; i.e., $f(X,Y) = kXY$. (f) Transactor symbol for which money flows in one direction and energy or matter in the other direction with price *(P)* adjusting one flow (J_1) in proportion to the other, $J_2(J_1 = PJ_2)$. (g) A combination of "active storage" and a "multiplier" by which potential energy stored in one or more sites in a subsystem is fed back to do work on the successful processing and work of that unit; autocatalytic. (h) Production and regeneration module *(P-R)* formed by combining a cycling receptor module, a self-maintaining module that it feeds, and a feedback loop that controls the inflow process by multiplicative and limiting actions; e.g., the green plant. (i) Sensor of the magnitude of flow, J.

because it influences the system. Mathematically, E can be some function of time, such as an increasing or decreasing linear function, a sinusoid, or a random signal. Thus $E = a + bt$, $E = a - bt$, and $E = a \sin(bt)$ are all possible functions for the source, E.

Constant Flow Source

This is a more precise description of a source indicating that an upper level, J_o, is available. The interaction symbol (described below) is meant to indicate that the production or output, J_2, is a function of X and what remains of the available energy, J_r. Thus, the equations describing this symbol are:

$$J_r = J_o - k_o J_r X$$
$$J_2 = k_1 J_r X$$
$$\text{or } J_2 = k_1 \frac{J_o X}{1 + k_o X}$$

representing a limiting relationship for the production, J, which is dependent on the flow of available energy, J_o. Thus, for example, X might represent nutrients in the system, and these may increase, but as X gets large, J_2 will approach a limiting value equal to $k_1 J_o / k_o$.

Heat Sink

In all real processes, energy is degraded from high-quality heat and is lost from the system as low-quality heat. This is a simple representation of the Second Law. Furthermore, solid lines in the system diagrams are used to represent flows of energy, matter, or information, although all can be theoretically given an energy value.

Storage (State Variable)

All systems accumulate storages of energy, matter, information, or money. Thus, storages of fuels, minerals, water, people, or money represent state variables in many systems. These storages can be added to or used up, and the basic differential equation for a storage is:

Rate of change of Q with time = inflow/time − outflow/time

$$\text{For Figure 3.3a: } \frac{dQ}{dt} = J - kQ$$

For each storage or state variable in a system, one differential equation can be written.

Interaction Symbol

This indicates that some output $f(x,y)$ is a function of two variables, x and y. Usually, it relates to production processes. Therefore, for example, photosyn-

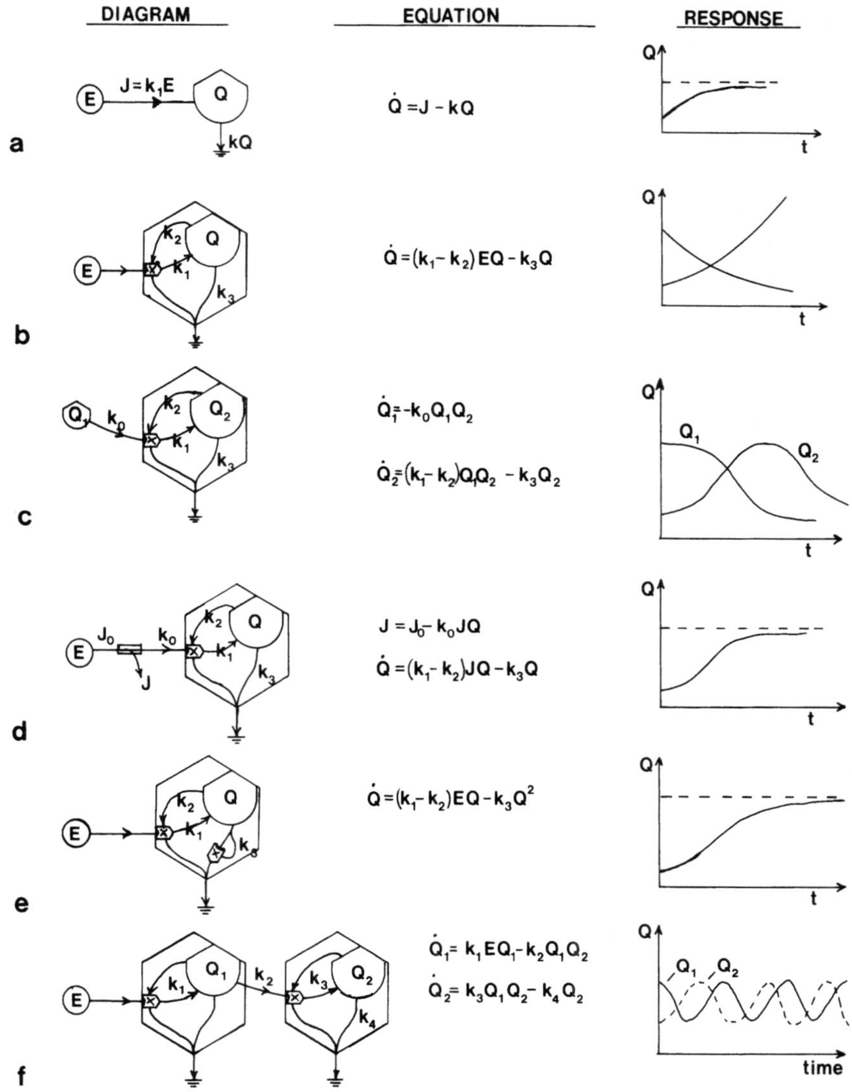

Figure 3.3. Some simple models and associated differential equations and expected responses. (a) Storage with constant inflow and outflow proportional to the amount of storage. (b) Autocatalytic growth. (c) Autocatalytic growth of storage (Q_2) off of a finite storage of energy (Q_1). (d) Autocatalytic growth of Q off of a constant flow source of energy. (e) Logistic growth of Q. (f) Predator-prey interaction model.

thetic production = k (nutrients) × sunlight; i.e., photosynthesis is a function of nutrients and sunlight. If several interaction symbols are in tandem, then production is a function of many variables. In economics, for example:

$$J = f(X,Y,Z) = kX^a Y^b Z^c$$

where J = output from an industry and X, Y, and Z are raw materials, capital, and labor, respectively. The heat sink on the interaction symbol is meant to indicate that there always are losses from the system during any interaction.

Price Transaction

This is to be used where there are money flows involved, and it indicates the exchange of money for energy, information, or matter. In general:

$$J_1 = \text{price} \times \text{quantity} = pJ_2$$

where price can be an arbitrary function of internal and external conditions.

Self-Maintaining Module

This is used to describe a system that must expend energy to capture energy for growth and maintenance. In ecosystems, it is usually reserved to describe heterotrophic organisms or populations. It consists of a storage that uses energy to capture external sources of energy; this is represented by the interaction symbol. For Figure 3.3b:

$$\frac{dQ}{dt} = (k_1 - k_2)\,\mathrm{E}\,\mathrm{Q} - k_3\,\mathrm{Q} - \text{output}$$

$$\frac{dQ}{dt} = \begin{matrix}\textit{(net}\\ \textit{growth)}\end{matrix} - \begin{matrix}\textit{(death and}\\ \textit{respiration)}\end{matrix} - \textit{(losses)}$$

This is usually used to indicate photosynthetic (autotrophic) units of an ecosystem with Michaelis-Menton kinetics associated with it. For example, if J is the photosynthetic production and I is the available light, then photosynthetic production would be described by the following relationship:

$$J = \frac{k_1 I}{k_2 + I}$$

There are many functional relationships that could be prescribed for photosynthetic production, and the exact form of the equations will indicate the assumptions.

Sensor

The small rectangle superimposed on a flow, J, is meant to indicate a reading of the level of J, which may trigger some action or decision in the system. For example, if the level of flow in a river goes beyond a certain level, a decision to ameliorate the flow may be made. Sensors are used to indicate information flows.

There are many variations and representations that can be generated for different systems (Odum 1983). The basic process is to first define the boundaries of a system of interest, to identify the storages or state variables, to specify the flows into and from the system and among the storages, to prescribe functional relationships for these flows, and to write a differential equation for each state variable. These models can then be used to describe how the system's state variables change over time and what the response will be to changes that occur internally or externally to the system. In order to generate realistic differential equations to describe a system, measurements of flows and storages must be made to parameterize a model. For example, if a flow, J, is assumed to be a product function of two state variables, Q_1 and Q_2,

$$J = K_o Q_1 Q_2$$

then measurements of J_1, Q_1, and Q_2 must be made to determine K_o. Knowing K_o then allows J to be calculated from new values of Q_1 and Q_2. Statistical methods, of course, can be used in this process.

In order to gain some familiarity with the process of generating model equations, the examples in Figure 3.3 are given, although detailed study of the models presented later in this volume can help in understanding the modeling process. The six basic models shown are as follows:

1. A storage with a constant inflow and outflow proportional to the size of the storage. The storage will approach a steady-state value, at which point inflow, J, equals outflow, kQ.
2. Exponential growth if $(k_1 - k_2) > k_3$; exponential decay if $k_3 > (k_1 - k_2)$.
3. Self-maintaining unit, Q_2, using a finite storage of energy, Q_1. Q_2 grows and eventually declines as Q_1 is exhausted.
4. Growth of Q, which is limited by the level of available flow of energy, J_o.
5. Logistic growth equation where outflow is a "self-interaction" term, k_3Q^2.
6. Predator-prey model, in which Q_2 feeds on Q_1 in proportion to the product of Q_1 and Q_2.

Theoretically, complex economic or ecologic systems could be described by a set of differential equations of the following form:

$$\frac{\partial Q_i}{\partial_t} = f(I_i; 0_i; Q_1, Q_2, \ldots, Q_n; \bar{X}; t) \quad i = 1,2,\ldots,n$$

where Q_i, i = 1,2,...,n are the state variables of the system.
I_i = inputs from the environment to storage, i.
O_i = outputs from the storage, i, to the environment.
$\bar{X}$ = space; i.e., in general, the model might have spatial dependence.
t = time.

This is a system of nonlinear differential equations for the n state variables. Given both data from the observable world to parameterize the model and a set of initial conditions for the state variables, Q_i, allows evaluation of the state of the system at any future time. Once a good representation of the system is arrived at, in the sense that the model output generates results corresponding to empiric measures, then the models can be used to anticipate future states of the system due to human or natural changes. Very simple models can be solved analytically, and these are sometimes useful for understanding general behavior. However, any system of nonlinear differential equations with more than two equations will need to be simulated and integrated numerically; computer programs are usually the means for accomplishing this task. Ecologic systems are almost always modeled with differential equations, because the rates of change of quantities within the system are large in comparison to the usual time frame of investigation; thus, one expects diurnal and seasonal changes in most of the state variables (Patten 1971–1976). Economic systems also change with time, but under many circumstances, the rate of change of a particular storage (e.g., capital plant) can be ignored during the period of interest. Under these conditions, the rate of change is assumed to be zero, and a differential equation approach is not required (this is explored somewhat in the previous section on "Input-Output Models"). In general, however, differential equation models based on fundamental laws of physical or biological systems would be sought as the most precise in terms of explanation, description, and prediction, assuming the systems under study can be described in terms adhering to fundamental laws.

Linear Programming and Optimization Models

A whole class of quantitative models can be specified that seek to maximize some objective function subject to constraints. Thus, one could have a function, $f(X_1, \ldots, X_n)$, m constraints, $g_1(X_1 \ldots, X_n)$, ..., $g_m(X_1, \ldots, X_n)$ that specify relationships among the variables; the objective might be to find the values of $X_1, \ldots, X_n$ that maximize or minimize the function, f (Courant 1936). The function, f, might be referred to as the objective function.

In many situations, it is possible to consider objective functions that are linear in terms of the variables (X's) as well as constraints, which are linear inequalities in terms of the X's (Intriligator 1971). Under these restrictions, use of linear programming models can be used in which the problem can be posed as follows:

maximize: $$V_1X_1 + V_2X_2 + \ldots + V_jX_j + \ldots + V_nX_n$$

subject to:
$$b_{11}X_1 + b_{12}X_2 + \ldots + b_{1n}X_n \leq r_1$$
$$\vdots$$
$$b_{m1}X_1 + b_{m2}X_2 + \ldots + b_{mn}X_n \leq r_m$$

and
$$X_1 \geq 0,\ X_2 \geq 0 \ \ldots,\ X_n \geq 0$$

where V_i represents the value per unit of X_i contributed to the objective function, b's specify the relationship of constraint i to variable X_j, and $r_1,\ldots, r_m$ are the values of the m constraints. For example, X_1 to X_n might represent the outputs of n industries, V_1 to V_n the value added per unit of output from each of the n industries, r_1 to r_m the constraints on available resources or pollution (e.g., b_1 is a constraint on available water, b_2 on energy, b_3 on labor, b_4 on pollutant output, and so on) and the coefficients b_{ij} the amount of resource i used per unit of output of industry j. Thus, the above problem would be to find the level of output of n industries so that the objective function of total value added is maximized subject to a set of resource constraints. The problem of minimization also can be considered and generalized as follows:

minimize: $$C_1X_1 + C_2X_2 + \ldots + C_nX_n$$

subject to:
$$d_{11}X_1 + d_{12}X_2 + \ldots + d_{1n}X_n \geq e_1$$
$$\vdots$$
$$d_{m1}X_1 + d_{m2}X_2 + \ldots + d_{mn}X_n \geq e_m$$

and $$X_1 \geq 0, X_2 \geq 0, \ldots, X_n \geq 0$$

Many different types of objective functions and constraints, of course, can be considered and optimization formulations could pertain to operations within a given sector or industry, as well as for a whole region of humans and nature. Possible objective functions at a regional scale might include maximization of total value added, total income, number of jobs, total work done, or environmental quality. Consideration of minimization problems might include such problems as minimizing total economic cost, total imports, environmental pollution, or total energy consumption. One interesting modification of the linear programming approach that incorporates the interdependency of the sectors included in a regional economy is to consider the input-output formulation of an economy. The output of the n sectors is related to final demand as follows:

$$\mathbf{X} = (\underline{\underline{I}} - \underline{\underline{A}})^{-1}\,\mathbf{Y}$$

The linear programming maximization problem then becomes (using matrix notation):

maximize: $$\mathbf{V}'\,\mathbf{X}$$

subject to: $$\underline{\underline{B}}\,\mathbf{X} \leq \mathbf{R}$$

and $$\mathbf{X} \geq 0$$

or maximize: $$\mathbf{V}'\,(\underline{\underline{I}} - \underline{\underline{A}})^{-1}\,\mathbf{Y}$$

subject to: $$\underline{\underline{B}}\,(\underline{\underline{I}}-\underline{\underline{A}})^{-1}\,\mathbf{Y} \leq \mathbf{R}$$

and $$\mathbf{X} \geq 0$$

Table 3.2. Intersectoral Flows for Gotland (in Million of Swedish Crowns, 1975)

To / From	Sector	Animal	Crops	1.2	1.3	2	3111	3112	3118	3.1	3.3
1.1	Animal	0	0	0	0	0	86	48.5	0	0	0
	Crops	30	2.5	0	0	0	0	0	20.7	11.9	0
1.2	Forestry	0	0	0.3	0	0	0	0	0	0	7
1.3	Fishery	0.5	0	0	0	0	0	0	0	0.8	0
2	Mining	0	0	0	0	0	0	0	0	0	0
3111	Slaughtery	0.4	0	0	0	0	0	0	0	1.9	0
3112	Dairy	0	0	0	0	0	0	0	0	0	0
3118	Sugar	0	0	0	0	0	0	0	0	0	0
3.1	Other food	42.7	0	0	0	0	0	0	0	0	0
3.3	Wood	0	0	0	0	0.7	0	0	0	0	0.3
3.4.2	Printing	0	0	0	0	0	0	0	0	0	0
3.6.9	Stone and soil	0	0	0	0	0	0	0	0	0	0.7
3.8.3	Communication equipment	0	0	0	0	0	0	0	0	0	0
Other workshops		5	2.4	0	0.3	0.9	0	1.6	0	0	0.3
4	Electricity and water	3.1	0	0	0	1.5	0.7	0.9	0.9	0.4	0.3
5	Construction	17.4	8.4	0	0	0	0	0	0	0	1.9
61/62	Wholesale	10.8	5.2	0	0	0	0	0	0	1.7	1
7	Transport	1.7	1	1.3	0	1.1	0	0	0	0	0
Tourism		0	0	0	0	0	0	0	0	0	0
8/9	Other services	0	0	0	0	0	0	0	0	0	0
Value added		103.8	45.2	12.4	2.1	27	29	15.6	15.5	14.2	17.2
Imports		0	39.1	4.2	3.1	0.5	5.8	1.7	2.1	20.4	0.9

where $\mathbf{V'}$ = row vector = $[V_1, V_2, \ldots, V_n]$.

$\mathbf{X}$ = column vector of sector outputs.

$$\underline{\underline{B}} = mxn \text{ matrix} = \begin{vmatrix} b_{11} & b_{12} & \ldots & b_{1n} \\ & \cdot & & \\ & \cdot\, b_{ij} & & \\ & \cdot & & \\ b_{m1} & b_{m2} & \ldots & b_{mn} \end{vmatrix}$$

$\mathbf{R}$ = row vector of resource constraints.

$\mathbf{Y}$ = column vector of final demands.

$(\underline{\underline{I}} - \underline{\underline{A}})^{-1}$ = Leontieff inverse.

Thus, the problem is transformed to finding a set of final demands, **Y**, so that the objective function is maximized. The attractiveness of the linear programming approach is that it is computationally tractable, it can handle a large number of variables, and—assuming that an objective function and constraints can be defined and good data can be made available—it is reasonably operational.

3.4.2	3.6.9	3.8.3	Other Workshops	4	5	61/62	7	Tourism	8/9	Total Output
0	0	0	0	0	0	0	0	0	0	214.8
0	0	0	0	0	0	0	0	0	0	103.8
0	0	0	0	0	0	0	0	0	0	18.2
0	0	0	0	0	0	0	0	1	0	5.5
0	3.9	0	0	0	0	0	0	0	0	31.7
0	0	0	0	0	0	0	0	1.6	0	121.5
0	0	0	0	0	0	0	0	0.4	0	68.3
0	0	0	0	0	0	0	0	0	0	39.2
0	0	0	0	0	0	0	0	1.4	0	51.3
0	0	0	0.3	0	2	0	0	0	14.6	29.6
0	0	0	0	0	0	8.5	0	0	0	17.5
0	0	0	0	0	1.3	0	0	0	0	130.9
0	0	0	0	0	0	0	0	0	0	237.6
0	0.7	0	3.7	0	0	0	0	0	0	39.2
0.3	12	1.3	0.5	1.2	0.5	2.4	0.7	1	0.4	28.1
0	0.7	0	0	1.2	0	0	0	0	0	135.6
0	0	0	0.5	0	0	1.8	0	64	0	256.9
0.3	0.1	0	0.1	0	4.8	27.6	18.4	1.4	18.9	153
0	0	0	0	0	0	0	0	0	0	96.7
0.7	0.1	0.1	0.9	0	4.8	20.8	24.1	0	17.7	165.4
15.3	67.3	189	23.4	7.4	103	163.6	78.3	14.3	78	
0.9	46.8	47.2	9.9	20.7	19.2	30.8	31.5	11.6	35.8	

Model 1: An Input-Output Model for Calculating Resource Use and Pollutant Generation

Important information for the purposes of regional planning or the completion of environmental impact analysis is the connection between various levels of economic activity, the consumption of resources, and the generation of pollution. Because of the complex nature of economies, a systems approach is needed to account for the various direct and indirect consequences of activity in one part of an economy. One rather simple and straightforward application of input-output analysis is for evaluating the direct and indirect consumption of resources per unit of output from a given economic sector. Again, because certain resources are in short supply or selected pollutants are critical to the environment, succinct measures for evaluating impacts for a given level of economic activity are needed. Andréasson (1984), using input-output data for 1975, both evaluated these measures by simplifying Gotland's economy down to 20 sectors and considered seven resources and pollutants that were of the most concern. The transaction flows among sectors were obtained by interviews with

Table 3.3. The Leontieff Inverse $(I-A)^{-1}$

To / From	Animal	Crops	1.2	1.3	2	3111	3112	3118	3.1	3.3
1.1	1.0066					0.7125	0.7148		0.0264	
Crops	0.1917	1.0247				0.1357	0.1361	0.5411	0.2427	
1.2	0.0012	0.0009	1.0171	0.0004	0.0075	0.0008	0.0009	0.0005	0.0007	0.2434
1.3	0.0054			1.0000		0.0038	0.0039		0.0157	
2					1.0000					0.0007
3111	0.0093					1.0066	0.0066		0.0372	
3112							1.0000			
3118								1.0000		
3.1	0.2002					0.1417	0.1422		1.0052	
3.3	0.0049	0.0031	0.0013	0.0018	0.0312	0.0035	0.0038	0.0021	0.0029	1.0124
3.4.2	0.0079	0.0058		0.0036	0.0001	0.0056	0.0056	0.0030	0.0073	0.0011
3.6.9	0.0011	0.0009			0.0008	0.0008	0.0008	0.0005	0.0003	0.0246
3.8.3										
Other workshops	0.0313	0.0263		0.0606	0.0413	0.0222	0.0482	0.0139	0.0079	0.0115
4	0.0221	0.0039	0.0005	0.0028	0.0660	0.0217	0.0299	0.0262	0.0129	0.0143
5	0.0984	0.0834	0.0001	0.0002	0.0050	0.0699	0.0705	0.0452	0.0226	0.0658
61/62	0.2401	0.1748		0.1107	0.0016	0.1699	0.1708	0.0923	0.2233	0.0346
7	0.0494	0.0397	0.0845	0.0157	0.1063	0.0350	0.0353	0.0210	0.0349	0.0283
Tourism										
8/9	0.0355	0.0271	0.0149	0.0146	0.0203	0.0252	0.0260	0.0144	0.0278	0.0119

To / From	3.4.2	3.6.9	3.8.3	Other Workshops	4	5	61/62	7	Tourism	8/9
1.1									0.0040	
Crops									0.0048	
1.2	0.0011	0.0003		0.0029	0.0002	0.0046	0.0025	0.0044	0.0014	0.0246
1.3									0.0159	
2		0.0300				0.0003				0.0001
3111									0.0007	
3112									0.0049	
3118										
3.1									0.0180	
3.3	0.0044	0.0013		0.0120	0.0009	0.0193	0.0105	0.0184	0.0056	0.1025
3.4.2	1.0000			0.0005			0.0331		0.0167	0.0001
3.6.9	0.0001	1.0001		0.0003	0.0005	0.0101	0.0003	0.0004	0.0002	0.0025
3.8.3			1.000							
Other workshops	0.0001	0.0072		1.1103		0.0003	0.0001	0.0002	0.0014	0.0012
4	0.0182	0.0982	0.0058	0.0161	1.0477	0.0054	0.0174	0.0064	0.0274	0.0050
5	0.0011	0.0099	0.0003	0.0015	0.0475	1.0015	0.0015	0.0015	0.0023	0.0068
61/62	0.0002	0.0001		0.0154		0.0007	1.0072	0.0006	0.5070	0.0035
7	0.0260	0.0048	0.0001	0.0099	0.0022	0.0470	0.1392	1.1636	0.0901	0.1517
Tourism									1.0000	
8/9	0.0494	0.0023	0.0005	0.0332	0.0023	0.0481	0.1173	0.2056	0.0627	1.1473

the major enterprises in each sector. This represented about 75–80% of total sales and accounted for the various industries, agriculture, and forestry. The service sectors were assumed to have proportionally the same transfers as for Sweden as a whole. Resource data were obtained as a result of the analyses of different sectors, as presented in Chapter 2. The economic flow matrix, **X**, eventually constructed, is contained in Table 3.2; the Leontieff inverse, $(\underline{\underline{I}} - \underline{\underline{A}})^{-1}$, is shown in Table 3.3. Total resource use in each of the sectors is presented in Table 3.4, and a matrix of resource requirements per unit of output is listed in Table 3.5. The resource requirement matrix explicitly presents the direct use of labor, oil, and water and the generation of biological oxygen demand, sulfur dioxide, and nitrogen oxides per unit of output from each of the sectors. Perusal of this matrix allows an immediate assessment of direct resource intensities for each of the sectors. One can also calculate the indirect effects by employing the input-output model. Thus, we can see that the total economic output is equal to the Leontieff inverse × the final demand vector (see the chapter on "Input-Output Models"):

$$\mathbf{X} = (\underline{\underline{I}} - \underline{\underline{A}})^{-1}\,\mathbf{Y}$$

The total resource use can be obtained from:

$$\mathbf{R} = \underline{\underline{B}}\mathbf{X}$$

where $\underline{\underline{B}}$ = $[mxn]$ matrix; each row entry, i, is the amount of direct resources, i, per unit of output from sector j.

R = $[mx1]$ vector of total direct resource requirements.

From the equation above

$$\mathbf{R} = \mathbf{B}(\underline{\underline{I}} - \underline{\underline{A}})^{-1}\mathbf{Y}$$

so that $\underline{\underline{B}}(\underline{\underline{I}} - \underline{\underline{A}})^{-1}$ represents the total resource requirements, direct and indirect, per unit of final demand.

The results of such analyses (Table 3.5) yield total resource use (direct and indirect) per unit of output, which can be compared to the direct use. For example, comparisons can be made between sectors as to resource or pollution generation per unit of output. As far as total impact, the oil intensiveness should be noted of such sectors as stone and soil, electricity and water, and the sugar industry in comparison to forestry, workshops, construction, or services. Activities such as agriculture, fisheries, and wholesale trade have high labor multipliers. With respect to water, sectors such as animal production, slaughter houses, dairies, and tourism have the highest requirements, while other sectors such as forestry and fisheries are quite low. Outlets of organic wastes create a high biological oxygen demand in animal production, slaughterhouses, and dairies. Emissions of sulfur dioxide are large in stone and soil, electricity production, and waterworks, while nitrogen oxides come from agriculture, forestry,

Table 3.4. Direct Resource Requirements and Pollutant Generation on Gotland, 1975

	Sector	Employment (No.)	Oil (TJ)	Electricity (TJ)	Water (1000 m^3)	BOD (tons)	SO_2 (tons)	NO_2 (tons)
1.1	Animal	2549		52	890	2830		
	Crops	1884	327				54.8	3129.4
1.2	Forestry	240	16		3		2.2	212.8
1.3	Fishery	116	41		1		5.7	545.3
2	Mining	175	51	35	2	1	12	3.1
3111	Slaughtery	427	28.3	12	179	55	6.7	1.2
3112	Dairy	251	112	15	198	130	26.4	4.7
3118	Sugar	175	215.4	18	57	3	50.7	9
3.1	Other food	203	69.3	8	96	42	16.3	2.7
3.3	Wood	433	14	7	6	3	3.4	0.6
3.4.2	Printing	208	6	6	20	2	1.4	0.3
3.6.9	Stone and soil	384	4050	351	338	3	956	168.5
3.8.3	Communication equipment	1249	33	30	110	4	8	1.4
	Other workshop industries	497	11	13	1	2	2.7	0.5
4	Electric and water	206	1633	26	19	1	384	67.8
5	Construction	1906		12	36	12		
61/62	Wholesale	2729	54	54	29	13	13	2.3
7	Transports	1533	430	16	15	7	36.1	4572
	Tourism	462	462	15	284	76	6	3336
8/9	Other services	2064	51	15	68	12	12	2.2
	Household		2993	391	4041	1060	528	5926
	Public sector	7233	363	77	83	103	86	15
	Totals	24,924	10,960	1153	6474	4359	2211.4	18,000.8

Table 3.5. Direct and Total Resource Requirements and Pollution Generation per Swedish Crown of Output, 1975

		Employment /1000 Skr		Oil (MJ)		Electricity (MJ)		Water dm^3	
	Sector	Direct	Total	Direct	Total	Direct	Total	Direct	Total
1.1	Animal	0.012	0.025	0.0	2.56	0.24	0.41	4.14	4.68
	Crop	0.018	0.024	3.15	3.69	0.0	0.09	0.0	0.08
1.2	Forestry	0.013	0.014	0.88	1.17	0.0	0.01	0.17	0.19
1.3	Fishery	0.021	0.025	7.46	7.74	0.0	0.07	0.18	0.22
2	Mining	0.006	0.009	1.61	6.04	1.10	1.20	0.06	0.14
3111	Slaughtery	0.004	0.021	0.23	2.41	0.10	0.40	1.47	4.78
3112	Dairy	0.004	0.022	1.64	4.34	0.22	0.54	2.90	6.23
3118	Sugar	0.005	0.018	5.50	8.94	0.46	0.53	1.45	1.51
3.1	Other food	0.004	0.015	1.35	3.24	0.16	0.27	1.87	2.13
3.3	Wood	0.015	0.021	0.47	2.43	0.24	0.35	0.24	0.39
3.4.2	Printing	0.012	0.013	0.34	1.56	0.34	0.37	1.14	1.18
3.6.9	Stone and soil	0.003	0.005	30.94	37.06	2.68	2.81	2.58	2.66
3.8.3	Communication equipment	0.005	0.005	0.14	0.50	0.13	0.14	0.46	0.47
	Other workshops	0.013	0.016	0.30	1.39	0.35	0.42	0.03	0.07
4	Electricity and water	0.008	0.014	61.62	64.58	0.98	1.03	0.72	0.77
5	Construction	0.014	0.015	0.0	0.81	0.09	0.14	0.27	0.33
61/62	Wholesale	0.020	0.024	0.34	1.87	0.34	0.40	0.18	0.30
7	Transports	0.010	0.015	2.80	3.74	0.10	0.15	0.10	0.21
	Tourism	0.006	0.019	5.69	7.86	0.18	0.41	3.49	3.72
8/9	Services	0.013	0.019	0.31	1.24	0.09	0.16	0.41	0.53

[a] BOD_N means output not connected to sewage treatment.

fisheries, sugar production, transportation, and tourism. It is interesting to note that the total value added per unit of output is rather evenly distributed over the sectors.

For convenience, Table 3.6 contains multipliers for the different sectors with regard to value added, primary resources, and pollutants. These multipliers indicate the total requirements for a given resource per direct resource consumption in a given sector. For example, for every person employed in the "slaughtery" sector 4.5 persons are totally employed in the economy. The ramifications of some of these results are discussed in Chapter 4.

Model 2: Regional Optimization Model Scenario Including the Effect of Energy Plantations

Because of primary interest in the effects of energy shortages on the economy and the possible contribution of renewable energy technologies, models were formulated to incorporate these two concerns. Subsequent to the first phase of the project, which entailed extensive analysis and data collection for the main activities on the island, an explorative optimization model for the total

BOD (g)		BOD_N[a] (g)		SO_2 (g)		NO_x (g)		Value Added	
Direct	Total	Direct	Total	Direct	Total	Direct	Total	Direct	Total
0.0	0.21	13.15	13.24		0.54		7.87	0.36	0.81
0.0	0.03			0.52	0.62	30.14	32.09	0.32	0.58
0.0	0.01			0.12	0.15	11.69	14.41	0.68	0.74
0.0	0.01			1.04	1.10	99.15	99.63	0.38	0.52
0.3	0.05			0.38	1.38	0.10	3.52	0.77	0.91
0.45	0.60		9.37	0.06	0.53	0.01	5.60	0.24	0.82
1.90	2.05		9.40	0.39	0.98	0.07	5.71	0.23	0.83
0.08	0.10			1.29	1.97	0.23	17.24	0.40	0.71
0.82	0.87		0.35	0.32	0.69	0.05	10.01	0.28	0.61
0.10	0.11			0.11	0.54	0.02	3.78	0.58	0.87
0.11	0.12			0.08	0.36	0.02	0.86	0.87	0.92
0.02	0.04			7.30	8.74	1.29	1.69	0.51	0.58
0.02	0.02			0.03	0.11	0.01	0.03	0.80	0.80
0.05	0.06			0.07	0.32	0.01	0.38	0.63	0.75
0.04	0.05			14.49	15.19	2.56	2.75	0.30	0.35
0.09	0.10				0.17		1.48	0.76	0.83
0.08	0.10			0.08	0.38	0.01	4.23	0.72	0.89
0.05	0.08			0.24	0.39	29.79	34.73	0.51	0.71
0.93	1.01		0.06	0.01	0.49	40.93	43.88	0.18	0.66
0.07	0.10			0.07	0.22	0.01	4.84	0.47	0.70

system was formulated. The emphasis was towards investigating the impact of energy constraints and the possibility of producing artificial oil from "energy plantation" forests. These energy plantations might consist of fast-growing poplar or willow trees, which would be harvested periodically. Bottom parts of the trees and root systems would remain and regrowth would take place. Fertilization and irrigation would be required to obtain economically acceptable yields. The biomass harvested could be processed into a liquid fuel as one alternative, although the wood fuel could be used in other ways. Details of this model and related general approaches can be found in Zucchetto et al. (1980b). Although by 1978 we had collected a great deal of energy, economic, and environmental data, there were still too many gaps for parameter evaluation of any detailed models. It is true that economic data on output by economic sector for 1972 had been collected, but sparse information existed for flows among sectors. Constraints on resources or pollutants were ignored in this model because of the great uncertainty concerning their evaluation. Even though there was much uncertainty with regard to precise numeric evaluation, the model was pursued as an "experiment" to gain possible insight to the response of the regional system to various scenarios. The intention was to give an illustrative example of possible future modeling approaches for our own and other projects

Table 3.6. Multipliers for Value Added and Primary Resources

Sector	Value Added	Employment	Oil	Electricity	Water	BO_7	SO_2	NO_X
Animal	1.35	1.67	—	1.41	1.09	—	—	—
Crop	1.05	1.17	1.10	—	—	—	1.12	1.05
Forestry	1.09	1.08	1.33	—	1.11	—	1.25	1.23
Fishery	1.11	1.05	1.01	—	1	—	1.01	1
Mining	1.11	1.33	3.08	1.07	1.83	1.25	3	20.2
Slaughtery	2.92	4.5	9.44	3.4	3.69	1.28	6	60
Dairy	3.09	4.75	2.49	2.18	2.37	1.05	2.39	72.4
Sugar	1.68	3.4	1.61	1.11	1.02	1.2	1.51	74.1
Other food	1.61	2.75	2.16	1.25	1.12	1.04	1.94	185.4
Wood	1.5	1.4	5.15	1.41	1.58	1.1	4.91	188.5
Printing	1.06	1.08	4.21	1.14	1.03	1.09	4	46
Stone and soil	1.14	1.67	1.2	1.05	1.03	1.5	1.14	1.28
Communication equipment	1	1	3.71	1.08	1	1	4	3
Other workshops	1.18	1.15	4.64	1.18	2	1.2	3.85	36
Electricity and water	1.25	1.75	1.05	1.05	1.07	1.25	1.05	1.07
Construction	1.09	1.14	—	1.56	1.3	1.11	—	—
Wholesale	1.27	1.36	8.24	1.24	1.85	1.4	7	4.24
Transports	1.39	1.5	1.34	1.5	2.22	1.6	1.63	1.17
Tourism	7.07	5	1.82	2.43	1.18	1.17	15.7	1.3
Services	1.5	1.45	3.87	1.78	1.28	1.43	2.7	4.84

as more precise data became available. The model formulation is essentially as follows:

$$\max_{(x,z)} \mathbf{V}_x\mathbf{X} + \mathbf{V}_z\mathbf{Z}$$

subject to:

$$\mathbf{X} \leqslant \mathbf{C}_x \quad \text{(production capacities)}$$
$$\mathbf{Z} \leqslant \mathbf{C}_z \quad \text{(land-use constraints)}$$
$$\underline{\underline{\mathbf{B}}}\,\mathbf{X} + \underline{\underline{\mathbf{D}}}\mathbf{Z} \leqslant \mathbf{R} \quad \text{(resource constraints)}$$
$$\underline{\underline{\mathbf{A}}}\mathbf{X} + \mathbf{Y} = \mathbf{X} \quad \text{(market equilibrium)}$$

since

$$\mathbf{X} = (\underline{\mathbf{I}} - \underline{\underline{\mathbf{A}}})^{-1}\,\mathbf{Y} \quad \text{the problem becomes}$$

$$\max_{(Y,Z)} \mathbf{V}_x(\underline{\mathbf{I}} - \underline{\underline{\mathbf{A}}})^{-1}\,\mathbf{Y} + \mathbf{V}_z\mathbf{Z}$$

subject to:

$$(\underline{\mathbf{I}} - \underline{\underline{\mathbf{A}}})^{-1}\,\mathbf{Y} \leqslant \mathbf{C}_x$$
$$\mathbf{Z} \leqslant \mathbf{C}_z$$
$$\underline{\underline{\mathbf{B}}}\,(\underline{\mathbf{I}} - \underline{\underline{\mathbf{A}}})^{-1}\,\mathbf{Y} + \underline{\underline{\mathbf{D}}}\,\mathbf{Z} \leqslant \mathbf{R}$$

where $\mathbf{X}$ = vector of output from economic sectors.
$\mathbf{Y}$ = vector of final demand (including exports).
$\mathbf{Z}$ = vector of forest and natural lands.
$\mathbf{R}$ = vector of resource availabilities.
$\underline{\underline{A}}$ = technical coefficients matrix of economic sectors.
$\underline{\underline{B}}$ = resource requirements coefficient matrix of the economic sectors.
$\underline{\underline{D}}$ = resource requirements coefficient matrix of the forest and natural lands.
$\mathbf{C}_x$ = vector of production capacities for economic sectors.
$\mathbf{C}_Z$ = vector of land-use constraints for forest and natural lands.
$\mathbf{V}_x$ = vector of value added coefficients for the economic sectors.
$\mathbf{V}_z$ = vector of value added coefficients for forest and natural lands.

Thus, the model maximized the total value added subject to production, resource, and market constraints. It included 13 economic sectors, as well as the natural ecosystem and forest land categories. To give some perspective on economic and resource levels associated with the model, the estimated resource requirements by sector, resource requirements coefficients, and the economic flow table for the 13 economic sectors are presented in Tables 3.7–3.9. The contribution from forest and natural lands is separated from the 12 main economic activities, because the introduction of energy plantations is considered to remove land from either of these categories. It is difficult to assign an appropriate economic value to the natural lands; in this model, we simply used the value added from tourist activities to generate a value from the natural system category. Resources required for these natural systems included labor, water, and land. The full optimization model included constraints on electricity, oil, gasoline, diesel fuel, imported goods, wood, water, labor, and land (Table 3.8). It was important to include such constraints on energy, because we were exploring a topic essential to the aims of the project; namely, the impact of shortages of energy on the island. Interruptions or shortages of energy could be simulated by modifying the constraint levels or the *R's*.

This model was simulated over time by postulating a number of scenarios that specified imported oil availability and oil price increases (Table 3.10). Reduction in oil availability is simulated by reducing the right-hand side of the resource constraint equation on oil. The increasing price of oil affects the objective function by reducing the value added vector, $\mathbf{V}_x$, by the cost of oil to each sector, while the value added vector, $\mathbf{V}_z$, is not affected because it was assumed that forest and natural lands did not need imported energy. In each year of simulation, the model is first run without energy plantations and then, with this sector, by introducing a value added term into the objective function and resource terms into the constraint equations. If the inclusion of artificially produced oil from energy plantations (scenario A oil production) enhanced total value added, then this situation was considered to be optimum; if not, the energy plantations were not included. It was assumed that the price of energy plantation oil was three times as high as the initial price of imported oil.

Table 3.7. Economic Flow Table for Model 2 (All Values in Millions of Swedish Crowns for 1972)[a]

From \ To	1	2	3	4	5	6	7
1. Stone and soil							
2. Food							
3. Quarries	2	1					
4. Workshops							
5. Graphics							
6. Wood	1		0.5				
7. Chemicals		0.3				0.3	
8. Textiles							
9. Agriculture		162.18					
10. Fisheries							
11. Electricity	0.95	0.222	0.173	0.117	0.02	0.026	0.026
12. Power works	3.75	0.878	0.687	0.463	0.08	0.104	0.104
Value added	53.2	97.5	13.703	91.474	7.398	9.295	2.374
Imports	26.4	16.62	6.937	0.246	0.902	6.175	0.596
Total gross outlay	87.3	278.7	22	92.3	8.4	15.9	3.1

[a] All blank entries are zeros.

Table 3.8. Resource Requirements by Sectors

	Types of Sectors							
Types of Resources	Stone and Sectors 1	Food 2	Quarries 3	Workshops 4	Graphics 5	Wood 6	Chemicals 7	Textiles 8
1. Electricity (TJ)								
2. Oil (TJ)	4024	334	13.3	1.74	4.8	5.5	0.93	1.7
3. Gasoline (TJ)								
4. Diesel (TJ)								
5. Goods (10^9 SKr)	13.78	15.52	6.87	0.24	0.87	6.14	0.59	0.59
6. Wood (TJ)						870		
7. Water (10^6 m^3)								
8. Labor (10^3 man-hr)	448	1094	210	1717	61	341	78	35
9. Land (ha)	1200	200	700	50	50	200	50	50

Table 3.9. Resource Requirement Coefficients (Amount Required/Unit of Output)

	Types of Sectors						
Types of Resources	Stone and Soil 1	Food 2	Quarries 3	Workshops 4	Graphics 5	Wood 6	Chemicals 7
1. Electricity (TJ)							
2. Oil (TJ)	46	1.20	0.604	0.019	0.571	0.346	0.30
3. Gasoline (TJ)							
4. Diesel (TJ)							
5. Goods (10^9 kr)	0.158	0.056	0.312	0.003	0.104	0.386	0.190
6. Wood (TJ)						54.7	
7. Water (10^6 m^3)							
8. Labor (10^3 man-hr)	5.12	3.92	9.54	18.6	7.26	21.4	25.2
9. Land (ha)	13.7	0.718	31.8	0.542	5.95	12.6	16.1

8	9	10	11	12	Final Demand: Local Demand	Final Demand: Exports	Output
					9.7	77.6	87.3
					75.3	203.4	278.7
					0.4	18.6	22
					17	75.3	92.3
					7.6	0.8	8.4
					8.2	6.2	15.9
					0.1	2.4	3.1
					3.2	0.1	3.3
					—	16.2	178.38
					—	5.66	5.66
0.0006					7.295	—	8.83
0.0024			4.75		12.772	—	23.59
2.695	21.92	5.08	4.08	2.5	—	—	311.22
0.602	156.46	0.58	—	21.09	50.92	—	287.528
3.3	178.38	5.66	8.83	23.59	192.5	406.2	726.86

Types of Sectors: Agriculture 9	Fishery 10	Electricity 11	Powerworks 12	Household 13	Forest 14	Natural land 15	Total Energy Available to Gotland
			544				544
1317			3112	2775			11,881
7.8	2.08			798.92			1024
239	35.57			445.43			720
							44.6
320							1190
90.33				12	7.6	7.3	300
4560	380	280	100	662	234	960	14,000
87,000	50	50	50	687	126,000	72,800	300,000

Types of Sectors: Textiles 8	Agriculture 9	Fishery 10	Electricity 11	Powerworks 12	Household 13	Forest 14	Natural land 15
				23.1			
0.515	7.38			132	14.4		
	0.0437	0.367			4.15		
	1.34	6.28			2.31		
0.179							
	1.79						
	0.506				0.0623	6.03×10^{-5}	10^{-4}
10.6	25.6	61.1	31.7	4.24	3.44	0.002	0.013
15.2	488	8.83	5.66	2.12	3.57	1	1

Table 3.10. Description of Scenarios[a]

Scenario	Energy Schedule	Time Horizon (Yr) 0	5	10	20
Scenario A	Annual rate of price increase	5%/yr	10%/yr	5%/yr	
	Imported oil availability	100%		80%	
	Artificial oil	Produced if beneficial			
Scenario B	Annual rate of price increase	5%/yr	20%/yr	5%/yr	
	Imported oil availability	100%		80%	
	Artificial oil	Produced if beneficial			

[a] Zucchetto et al. (1980b).

Results for two scenarios (Table 3.10) are illustrated in Figures 3.4 and 3.5. The price of oil is allowed to rise in both scenarios at 5%/yr for the first 5 years, and then a sudden jump takes place to 10%/yr for scenario A and 20%/yr for scenario B. When this jump takes place, the constraint on oil imports is set at 80% of the initial value. Smaller price increases of 5%/yr are then assumed for the 10th to 20th years. If artificial oil is introduced, the constraint on labor is relaxed by 20% because of the enhanced labor demand. The results of these two simulated scenarios show that agriculture, food, workshops, chemical, and textile sectors remain quite insensitive to oil restrictions and price increases under the objective function of maximizing value added. This is due to the fact that these sectors generate relatively high levels of value added per unit of energy input. However, energy-intensive sectors such as stone and soil and quarries are significantly affected. In both scenarios, large increases occur in the labor employed and wood used. The model chooses to introduce artificial oil production in the eighth year for scenario A and in the sixth year for scenario B, both of which correspond to times for which the price of artificial oil is still more than the price of imported oil.

Model 3: A Regional Optimization for 1975 with Emphasis on Water Constraints and Energy Prices

One of the last models to be considered in the project was a 20-sector economic optimization model (Andréasson 1984)—data that have already been presented under model 1 (see Tables 3.2–3.5). This model is somewhat different than model 2 in terms of its analysis because of the lack of the time element, but it goes beyond model 2 because there are a greater number of sectors, the economic data base is better, a more recent year (1975) is considered, and resource use and pollutant data are included and updated. The emphasis in the present model is on the effect of water resources and energy prices on regional economic activity. Also included were exports, imports, and resource use associated with

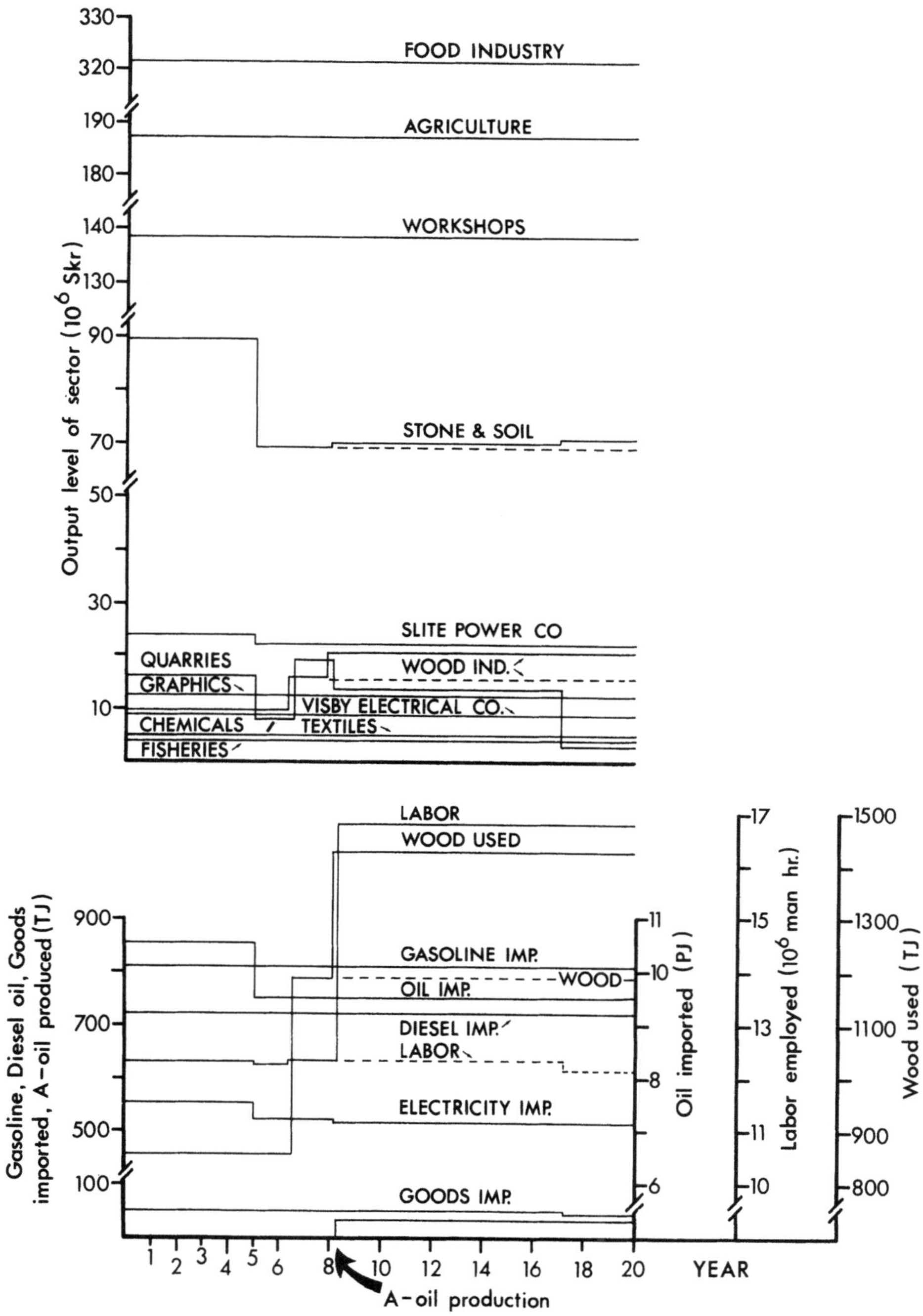

Figure 3.4. Model results for scenario A in Table 3.10. The upper graph represents the output level of different activities, while the lower graph refers to resource use. Dotted lines refer to results without artificial oil production (Zucchetto et al. 1980b).

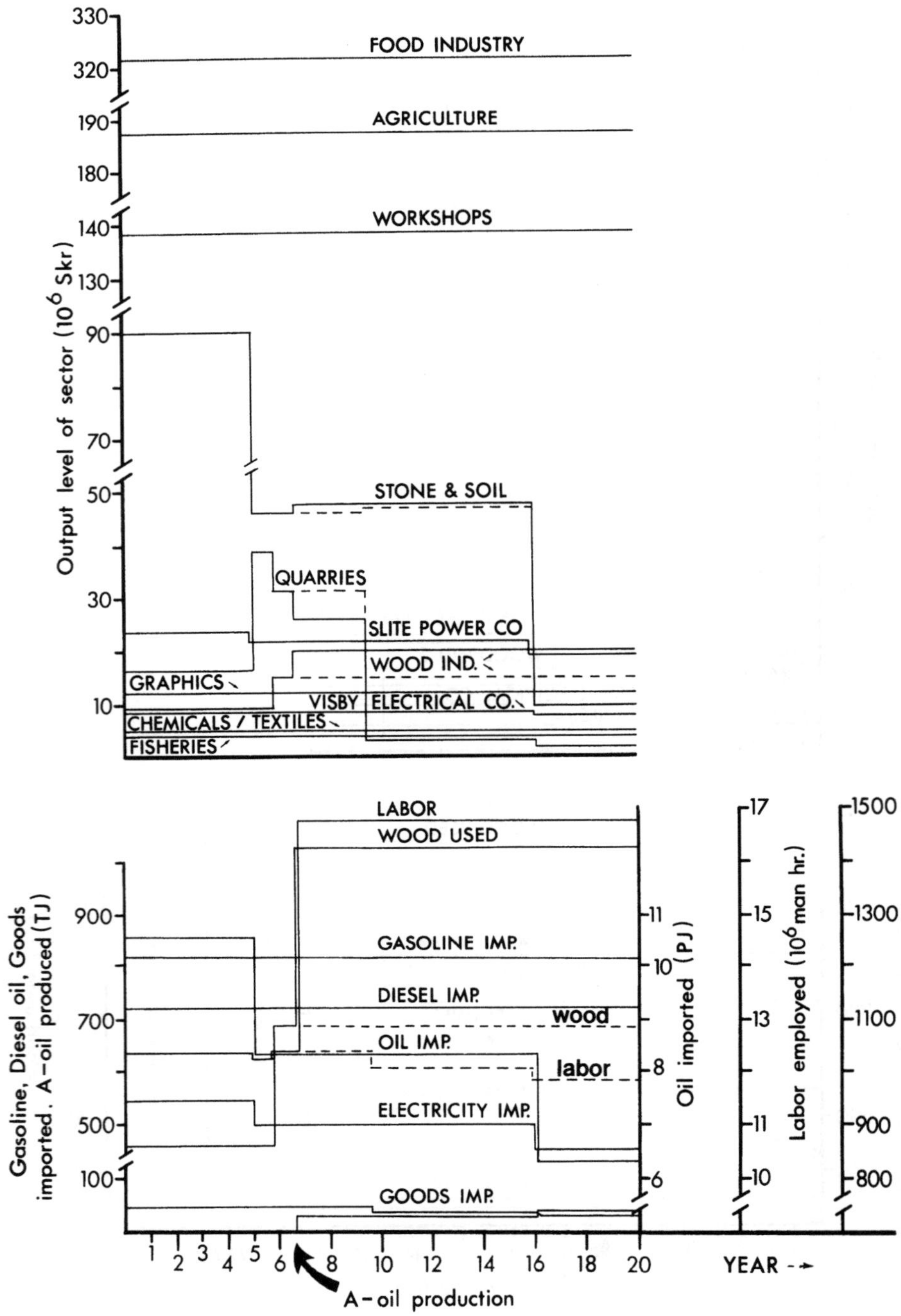

Figure 3.5. Model results for scenario B in Table 3.10. The upper graph represents the output level of different activities, while the lower graph refers to resource use. Dotted lines refer to results without artificial oil production (Zucchetto et al. 1980b).

the household sector. The formulation of the model was essentially a linear programming one in terms of the sector outputs, X_i's.

The objective function was to maximize value added associated with production minus value added attributed to imports to each sector:

$$\max_{x,m,e} (\mathbf{VX} - \mathbf{tM})$$

where $\mathbf{V}$ = $1 \times n$ vector of value added per unit of output from each sector i.
$\mathbf{X}$ = $1 \times n$ vector of outputs from each sector.
$\mathbf{t}$ = $1 \times n$ vector of value added associated with imports to each sector i.
$\mathbf{M}$ = $n \times n$ diagonal matrix whose elements are imports to each sector.

The constraints on the problem were as follows:

Equilibrium:

$$(\underline{\underline{I}} - \underline{\underline{A}})\,\mathbf{X} + \mathbf{m} - \mathbf{zC} - \mathbf{e} = \mathbf{g} + \mathbf{i}$$

where $\underline{\underline{I}}$ = $n \times n$ identity matrix.
$\underline{\underline{A}}$ = $n \times n$ coefficient matrix.
$\mathbf{X}$ = $n \times 1$ vector of sector outputs.
$\mathbf{m}$ = $n \times 1$ vector of imports to each sector.
$\mathbf{C}$ = $n \times 1$ vector of consumption; each element of which = C_i = $z_i c$, where z_i is the proportion of total consumption, c, for each sector. Total consumption $c = p \cdot \mathbf{v} \cdot \mathbf{X}$, where p is the same as for Sweden as a whole.
$\mathbf{e}$ = vector of exports.
$\mathbf{g}$ = vector of government consumption.
$\mathbf{i}$ = vector of investment.

Resources:

$$\underline{\underline{B}}\,X + \mathbf{JC} \leq \mathbf{r} - \mathbf{Dg}$$

or

$$\underline{\underline{B}}X + \mathbf{JC} + \mathbf{Dg} \leq \mathbf{r}$$

where $\underline{\underline{B}}$ = matrix of resource use per unit of output from each sector.
$\mathbf{J}$ = vector of resource consumption per kronor of consumption in each sector.
$\mathbf{r}$ = capacity limit on resource consumption.
$\mathbf{D}$ = resource consumption per unit of consumption in government sector.

Also, it was assumed that consumption is proportional to income (see above):

$$c = p \cdot \mathbf{v} \cdot \mathbf{X}$$

Capacity constraints were also set on output from each sector:

$$\mathbf{X} \leqslant \mathbf{k}$$

where $\mathbf{k}$ = vector of capacity constraints.

$$\mathbf{X}, \mathbf{m}, \mathbf{e} \geqslant 0$$

Other constraints included employment, with an upper limit on labor available and a lower limit equal to the employment in 1975 (22,000–27,000 people). The highest level of fertilization permitted in agriculture was derived from the water-nitrogen model (model 6 in this chapter) so as to maintain groundwater nitrate levels below 8 mg/liter, implying that the total acceptable use of fertilizer was less than 97 kg N/ha $\times$ 83,000 ha = 8051 tons of nitrogen (N). This is one example of a coupling between an economic model and an environmental model. In this case, the environmental model was used to derive an acceptable level of the constraint function for the economic model. The biological oxygen demand (BOD) constraint was set by the capacity limits of the sewage treatment plants (BOD $\leqslant$ 1372 tons). For households not connected to the sewage treatment plants, the upper limit was set at 3180 tons. Sulfur dioxides and nitrogen oxides were permitted to increase 10 times above the 1975 level; total $SO_2 \leqslant 23{,}600$ tons and total $NO_X \leqslant 180{,}000$ tons. The total amount of oil was restricted by the storage and transportation capacity (oil $\leqslant$ 15,200 TJ).

Since water is a particularly critical resource on the island, the simulation focused on the effect of varying the water constraint to determine its effect on the value of the objective function. Simulations were also conducted for changes in the BOD constraint and energy prices. Table 3.11 summarizes household consumption, intermediate flows, investment, exports, production levels, imports, taxes, and production constraints for 1975. The effect of water is summarized in Table 3.12 and Figure 3.6. As the water constraint is relaxed, production increases, but in a diminishing returns fashion; beyond about 7 million m^3 per year available for economic production, no increase in the gross regional product (GRP) occurs. At low levels of water availability, activity in certain sectors is virtually eliminated (Table 3.12). The ideal market-determined price or shadow price for water varies between 0.022 and 0.276 Skr liter^{-1}, which is high in comparison to the actual price of 0.0012 Skr liter^{-1} for 1975. The effect of raising water prices on the price of final goods from each sector is summarized in Table 3.13 and is compared to the effect of changes in oil prices. The rise in consumer prices is relatively small, even for significant increases in water prices, while increasing oil prices have a much more severe impact. These calculations were made with input-output analysis.

Simulations were also conducted for the effect of changing the BOD constraint

Table 3.11. Final Demand and Capacity Limits (in Millions of Swedish Crowns, 1975)

Sector	C_i Household	Interme-diate Flows	I_i Investment	E_i Export	G_i Gov't. Investment	X_i Production	M_i Import	Taxes	Capacity Limit
Animal	4.7	134.5		43.3	0.6	214.8		31.7	252.7
Crops	23.9	65.1			3.3	103.8	7.7	19.2	142.2
Forestry		7.3	0.8	8.3	1.9	18.2		0.1	24.6
Fishery	0.1	2.3		3.7		5.5		0.6	9.6
Mining		3.9		28.7		31.7		0.9	43
Slaughtery	49	3.9		109.4	3.4	121.5	33.1	11.1	178.4
Dairy	28.6	0.4		54.6	1.8	68.3	9.4	7.7	100.3
Sugar	4.1			39.2		39.2	3.7	0.4	57.6
Other food	100.2	44.1			2.9	51.3	90.4	5.5	75.3
Wood	9.9	17.9			5.1	29.6	3.2	0.1	35
Printing	18.9	5.3			15.9	17.5	21.7	0.9	31.1
Stone and soil		2		128.9		130.9			146.6
Communication workshops	1.1		3.1	242.1	1.8	237.6		10.5	283.3
Other workshops	338.7 +1.8	14.8	61		31.8	37.3	409.7	1.1	45.5
Electric and water	17.2	28.2			3.4	28.5	22.2	1.9	53
Construction		29.6	84.1	0.5	27.4	135.6		6	183.7
Wholesale		106.4		45.5	9.1	161			
Transports	22.9	67.5		46.9	16.2	153.5			
Tourism	41.4			57.1	1.7	81.5	18.7		110
Other services	228.8	61.5	7.4		55.9	165.4	184.1	4.1	
Oil	62.7	66			3.6		132.3		

Table 3.12. Optimization Results for Different Water Constraints (in Millions of Swedish Crowns, 1975)

Sector	Water $\geq 8 \times 10^6$ m^3			Water $\leq 4.5 \times 10^6$ m^3			Water $= 6.5 \times 10^6$ m^3 [a]		
	Prod.	Export	Import	Prod.	Export	Import	Prod.	Export	Import
Animal	239.3	36.9					152.6		50.4
Crops	35.2		74.9	136.5	106		132.9	33.3	
Forestry	24.5	13.1		24.5	131.4		24.5	13.1	
Fishery	9.6	6.2		9.6	9.5		9.6	8	
Mining	43	38.6		43	43		43	38.6	
Slaughtery	178.4	120.3				53.5	178.4	122.9	
Dairy	100.3	69.2				31.1	100.3	69.8	
Sugar	57.6	53.5				42	57.6	53.5	
Other food	75.3		79.1			105.2	75.3		59.3
Wood	35		13.8	35		1.4	35		13.8
Printing	31.1		14.8			46.2	31.1		14.8
Stone and soil	146.6	144				3.2	146.6		
Communication workshops	283.3	277.3		211.5	205.5		283.3	277.3	
Other workshops	45.5		684.6	45.5		661.8	45.5		718
Electric and water	53.6		1			26.2	53.6	1.7	
Construction	183.7	44.5		183.7	58.9		183.7	43.6	
Wholesale	3.31	194.7		331	293.8		331	255.5	
Transport	310			310	164.7		310	162.7	
Tourism	126	83.2				52.9			52.9
Other services	331		80.8	329.7		80.6	331		80.8

[a] Actual use, 1975.

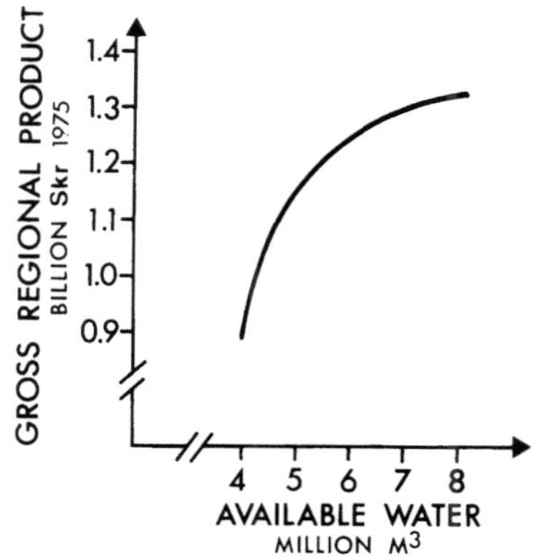

Figure 3.6. Results of maximizing value added and changing the available water constraint. Relaxing the water constraint results in greater regional product, but at a diminishing rate. The regional product is in constant Swedish crowns (Skr) at 1975 prices (Andréasson et al. 1983).

Table 3.13. Percent Increase in Price of Final Goods with Respect to Changes in Water and Oil Prices, 1975[a]

Sector	1	2	3
Animal	10	5	2.9
Crops	0.2	0.1	4.5
Forestry	0.4	0.2	1.9
Fishery	0.6	0.3	10.1
Mining	1.2	0.6	7.7
Slaughtery	10.2	5.1	2.7
Dairy	13.2	6.6	8.2
Sugar	3.4	1.7	10
Other food	4.6	2.3	3.8
Wood	0.8	0.4	2.3
Printing	2.6	1.3	1.1
Stone and soil	6.8	3.4	38.7
Communication workshops	1	0.5	0.5
Other workshops	0.4	0.2	1.2
Electric and water	15.2	7.6	64.8
Construction	0.2	0.1	1
Wholesale	0.4	0.2	2.1
Trade	0.4	0.2	9
Tourism	7.8	3.9	25.8
Other services	1.2	0.6	1.3

[a] Column 1. The percentage increase in the price for final goods when water price increases 20 times. Column 2. The same as column 1 for a 10-times increase in water prices. Column 3. The same as column 1 for a doubling of the oil price.

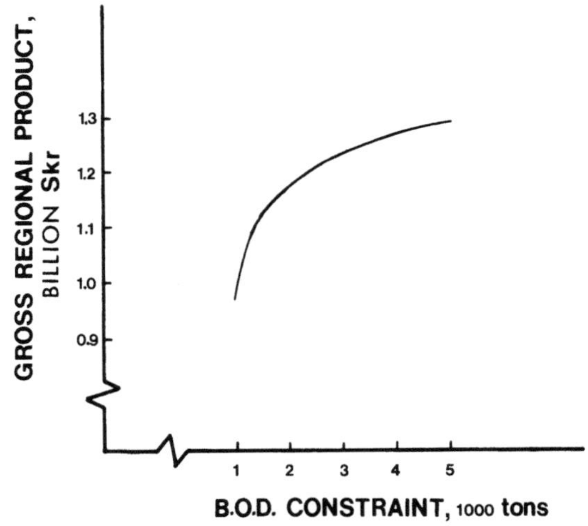

Figure 3.7. Effect of changing the BOD constraint on the optimum level of the GRP (Andréasson 1984).

on the GRP. The impact is small until the constraint is severely reduced to about 1400 tons (Figure 3.7). At high levels of BOD (4000 tons), the shadow price is about 0.024 Skr g^{-1}; whereas at 1300 tons, it is 0.27 Skr g^{-1} and at 1000 tons, 5.8 Skr g^{-1}. Although not presented in detail here, optimization results were also obtained for changed energy prices and transport subsidies. In comparison to the base case (Figure 3.6), an increased electricity price of 25% reduced GRP by about 2%, an oil price increase of 25% reduced GRP by about 2%, and an oil price decrease of 25% increased GRP by some 1.5%. Elimination of the transport subsidy had little impact on the optimal level of GRP, reducing it by only 2%. There was little impact on total employment, with an oil price rise of 25%. The transport subsidy helps to make goods from Gotland more competitive. The GRP was sensitive to the prices of important exports. Reduction of the price of cement by 25% reduced GRP by 2.5%; lowering the price of animal and dairy products by 25% lowered GRP by 6%. These results are approximate measures, because they vary depending on the water constraint assumed.

Model 4: Energy Diversity and Stability of the Economic System

Research into the effect of energy on the economy led to considering the impact that diversification might have in enhancing the stability of the economic system. In the ecologic literature, the relationship of species diversity and the stability of ecosystems has been studied in several ways (Connell and Orias 1964, E.P. Odum 1971, May 1973, H.T. Odum 1973, Pielou 1975, Odum and Odum 1976, 1981). Stability in this literature usually refers to some notion of an ecosystem's ability to recover its original behavior after being subjected to some perturbation.

Table 3.14. Energy Source-Activity Matrix

	Sector (Activity)				
Energy Source	1	2	...	m	
1. Oil	e_{11}	e_{12}	...	e_{1m}	$E_1\bullet$
2. Gasoline	e_{21}	e_{22}	...	e_{2m}	$E_2\bullet$
3. Coal	·				·
4. Oil-electric	·				·
5. Wind-electric	·				·
6. Wood-electric					
·			e_{ij}		
·					
·					
(n) ...	e_{n1}	e_{n2}	...	e_{nm}	$E_n\bullet$
	$E_1\bullet$	$E_2\bullet$	...	$E_m\bullet$	E

A system that is able to recover from perturbations is thought to have a high stability. Although strongly criticized, many ecologists have hypothesized that higher diversity leads to greater stability. For the purposes of pursuing this line of thought, with regard to energy flow systems, a formulation by Watt (1972) seems most appropriate. In this formulation, he discusses the diversity of food (energy) sources to a number of species in a matrix, with different types of food represented by the rows and different species along the columns; each entry of the matrix, F_{ij}, represents the amount of food of type "i" eaten by species "j." Greater stability is attributed to the situation in which the diversity of food sources is greater. A variation of Watt's approach has been used for regional economies by Zucchetto (1981), in which an energy source-activity matrix was defined (Table 3.14), where each entry, e_{ij}, refers to the amount of energy from source "i" used in economic sector "j." There are various theoretic aspects that can be investigated by means of such a model. For example, it might be desirable to maximize the diversity of energy sources, because this might result in greater stability of an energy system. However, diversification engenders a cost, either in dollar or energy terms, that must be taken into account; an objective incorporating cost might be to maximize diversity per unit cost. One possible formulation is as follows:

$$\max h = \frac{H}{C} = -\frac{\Sigma X_{ij} \log_e X_{ij}}{E \cdot \Sigma P_{ij} X_{ij}}$$

subject to $E \cdot \Sigma P_{ij} X_{ij} = C =$ total cost and $\Sigma X_{ij} = 1$

where H = diversity index = $-\Sigma_{i,j} (X_{ij} \log_e X_{ij})$.

X_{ij} = e_{ij}/E = fraction of total energy E of type "i" used in economic sector "j."

P_{ij} = cost of energy type "i" used in activity "j."

E = total energy used by system.

Table 3.15. Flow Matrix Used for the Energy Diversity Model (Flows are in Millions of Swedish Crowns for 1975). The Energy Source Matrix for Two Sources is Presented for Illustration

Sector	1	2	3	4	5	6	7	8	Final Demand	Total Output
1. Agriculture	102	0	191	0	0.3	0	0	0.9	20.1	314.3
2. Quarries	0	0	0	3.9	0	0	0	0	26.8	30.7
3. Food industry	1.9	0	4.5	0	0.3	0	0	11.7	265.8	284.2
4. Stone and soil	0	0	0.3	1.2	0	0	0	17	102.6	121.1
5. Other industries	0	0.9	0	0.9	13.6	0	0	27.4	281	323.8
6. Transportation	8.2	0	0	0	0.3	25.7	12.8	22	128.4	197.4
7. Tourism	0	0	0	0	0	0	1	0	165.8	166.8
8. Other services	6.8	0	0	0.7	22.7	57.3	67.2	67.9	225.1	447.7
Value added plus imports	195.4	29.8	88.4	114.4	286.6	114.4	85.8	300.8		
Total outlay	314.3	30.7	284.2	121.1	323.8	197.4	166.8	447.7		1886
Energy matrix[a]										
Fossil fuels (TJ)	327	51	425	4050	65	736	344	105		
Electricity (TJ)	52	35	53	351	53	16	15	85		
Total	379	86	478	4401	118	752	359	190		

[a] Example for two energy Sources (see Table 3.16).

The solution to this problem is:

$$\frac{X_{ij}}{X_{kl}} = e^{A(P_{kl} - P_{ij})}$$

where $A = \lambda CE$ (λ = Lagrange multiplier)

which means that the ratio of two energy entries in the energy source-activity matrix is exponentially related to the difference in cost. Results such as these could be compared to existing conditions to determine whether a given region is evolving in such a way as to maximize a given objective function. Other constraints and considerations could be introduced; however, for the purposes of analyzing an actual economic system, more realism would probably be achieved by the formulation of simulation models.

A simulation using realistic data was conducted to deal with the effect of energy diversity on economic activity for Gotland. Recalling the input-controlled formulation discussed in the previous section on "Input-Output Models," we have the following:

$$\mathbf{X}' = \mathbf{V}' (\mathbf{I} - \underline{\underline{\bar{\mathbf{A}}}})^{-1}$$

where $\mathbf{X}' = [X_i]$ = row vector of output from each sector.
$\mathbf{V}' = [V_i]$ = row vector of value added and imports.
$\underline{\underline{\bar{\mathbf{A}}}} = [\bar{a}_{ij}]$ = matrix of input-controlled technical coefficients where $\bar{a}_{ij} = X_{ij}/X_i$, X_{ij} = transaction from sector "i" to "j."
$\underline{\underline{\mathbf{I}}}$ = identity matrix.

Thus, knowing the value added and the structure of the economy permits the calculation of sector outputs. The flow table used for 1975 is shown in Table 3.15. In order to incorporate the energy source-activity matrix, both the number of energy sources and the amount of energy of each type used by each economic sector were stipulated; the five energy source-activity matrices for which simulation experiments were conducted are delineated in Table 3.16.

The manner in which simulation experiments were conducted is described as follows. A maximum percent interruption for each energy source was stipulated and a large number of runs were conducted, with interruption of each energy source chosen randomly between zero and the maximum specified. Each energy source was independent of the others. Once the percent interruption was randomly chosen, it was assumed that value added was affected in the same way; this allowed calculation of the total output, X. Simulations were conducted for five energy source-activity matrices (Table 3.16). Simulations showed a reduction in variability of output as the energy system diversified (Figures 3.8 and 3.9). Simulations were also conducted with the incorporation of a priority scheme by which energy available was allocated according to the energy efficiency of the economic sectors; those sectors that produced more economic output per unit of energy consumed were given priority in the use

Table 3.16. Energy Source Activity Matrices Used for the Simulations (TJ/yr)

	Sector							
	1	2	3	4	5	6	7	8
Source								
1. Fossil fuels	327	51	425	4050	65	736	344	105
2. Imported electricity	52	35	53	351	53	16	15	85
Total	379	86	478	4401	118	752	359	190
Source								
1. Fossil fuels	327	51	225	3650	15	736	344	105
2. Imported electricity	52	35	53	351	53	16	15	85
3. Wind-electrical	0	0	200	400	50	0	0	0
Total	379	86	478	4401	118	752	359	190
Source								
1. Fossil fuels	127	51	125	3150	15	661	269	105
2. Imported electricity	52	35	53	351	53	16	15	85
3. Wind-electrical	0	0	200	400	50	0	0	0
4. Biomass	200	0	100	500	0	75	75	0
Total	379	86	478	4401	118	752	359	190
Source								
1. Fossil fuels	127	51	75	3100	15	661	269	105
2. Imported electricity	52	35	53	351	8	16	15	85
3. Wind-electrical	0	0	200	400	50	0	0	0
4. Biomass	200	0	100	500	0	75	75	0
5. Wave-electrical	0	0	50	50	45	0	0	0
Total	379	86	478	4401	118	752	359	190

of available energy. There are many ways of simulating this procedure, one of which is to use a parameter, b, which reduces the interruption of energy to the energy-efficient sectors so that the interruption is $b \times$ what was randomly chosen ($o \leqslant b \leqslant 1$). For one set of simulation runs, we considered sector 4 (stone and soil) as being the only energy-inefficient sector and all other sectors as being energy-efficient. For example, for a value of $b = 0.5$, the level of the energy disruption to all sectors is reduced by 50% at the expense of energy supply to sector 4. Simulations were conducted with two and five energy sources and with reductions of energy interruptions to the energy-efficient sectors of 25% and 100%, respectively (Figures 3.10–3.13). There is a significant reduction in the variability of total economic output as energy sources diversify and a priority allocation is incorporated.

As in most models, this one entails uncertainties. One must determine technical coefficients and calculate an energy source-activity matrix based on the types of energy sources anticipated with reasonable approximations as to how much energy can be produced and how it is to be distributed among sectors. Then, the structure of the economy must be ascertained based on economic data, preferably derived from as recent a year as possible. Next, one has to prescribe probabilities of interruption and different allocation schemes for energy

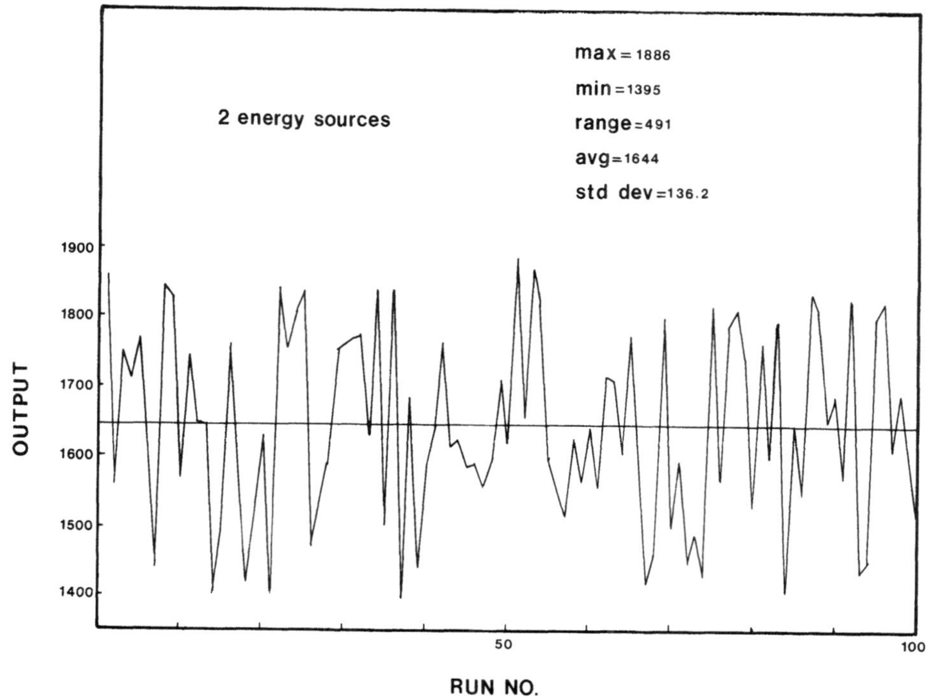

Figure 3.8. Total output of Gotland's economic system (in millions of Swedish crowns for 1975) for 100 simulation trails. Two energy sources (fossil fuels and imported electricity) were used by the eight economic sectors. Fossil fuels were assumed to have a maximum probability of interruption of 30% and imported electricity, 15%.

among sectors. If this can be accomplished, the model can generate a range of responses to several energy supply alternatives.

Model 5: Optimization Model for Gotland's Energy Supply System

In previous chapters, detailed data of the type and amount of energy used in different activities on Gotland have been presented. The potential use of alternative energy technologies for meeting future energy demand has also been mentioned. In the present section, attention is directed towards optimization approaches with regard to planning an energy supply system for meeting the needs of the local economy. Detailed assumptions and calculations of this work can be found in Ahlbom (1982). The optimization model includes both renewable and nonrenewable energy supplies, as well as consideration of various physical, environmental, and economic constraints.

Two major issues associated with the energy needs of the Gotland economy were investigated with this optimization model. As discussed repeatedly in this

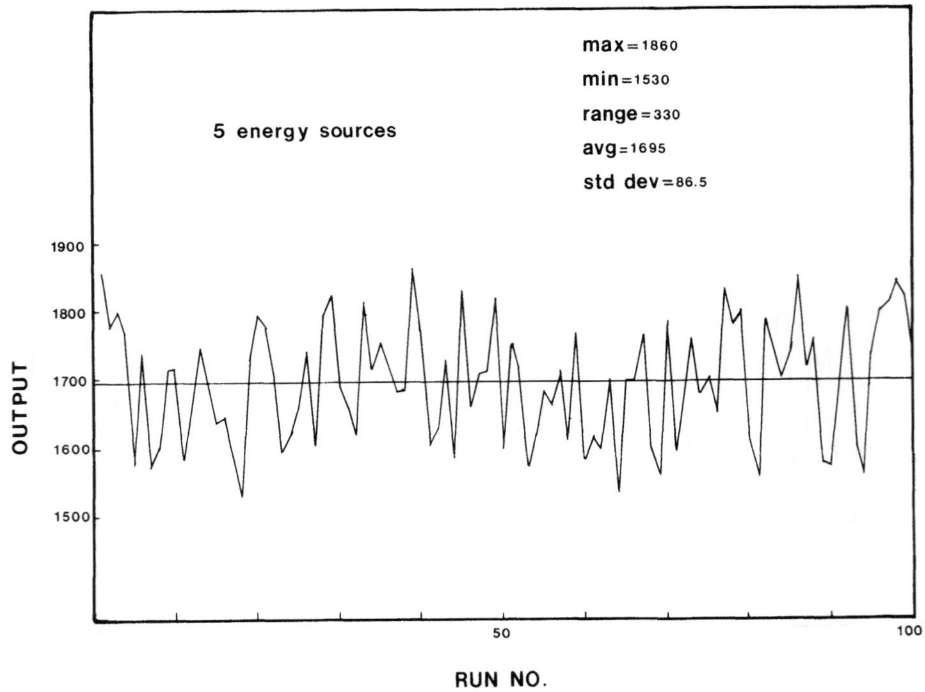

Figure 3.9. Total output of Gotland's economic system (in millions of Swedish crowns for 1975) for 100 simulation runs. Five energy sources (fossil fuels, imported electricity, wind-electric, biomass, and wave-electric) were used by the eight economic sectors. Maximum probabilities of interruption were: (1) fossil fuels, 30%; (2) imported electricity, 15%; (3) wind-electric, 20%; (4) biomass, 0%; and (5) wave-electric, 20%.

volume thus far, the island of Gotland is dependent on external sources of fuels and electricity for activities in the economic system. Our investigations, as well as those of Kjellström and Gustafsson (1976), Anonymous (1977b), Anonymous (1979c), Eneroth et al. (1979) and Ljungblom et al. (1978), have tentatively suggested and estimated the contributions made by renewable energy technologies to Gotland's economy. The present modeling effort includes alternative energy technologies that supposedly could contribute to the economy within 10 years; it includes straw, wood wastes, peat, energy (trees) plantations, heat pumps, solar collectors, and wind-electrical and wave-electrical energy sources. One issue, then, is to estimate the maximum contribution that local energy sources could contribute. The provision of energy from a given source entails a certain economic cost that depends on many factors. Thus, it might be conceivable to supply a given amount of energy from a local energy source, but at costs that might be socially unacceptable because of its impact on other parts of the economy. To this end, optimization models were also formulated to determine the mix of energy sources, with the objective of minimizing total costs of the energy supply system.

These models essentially focus on the local constraints of wind, water, suit-

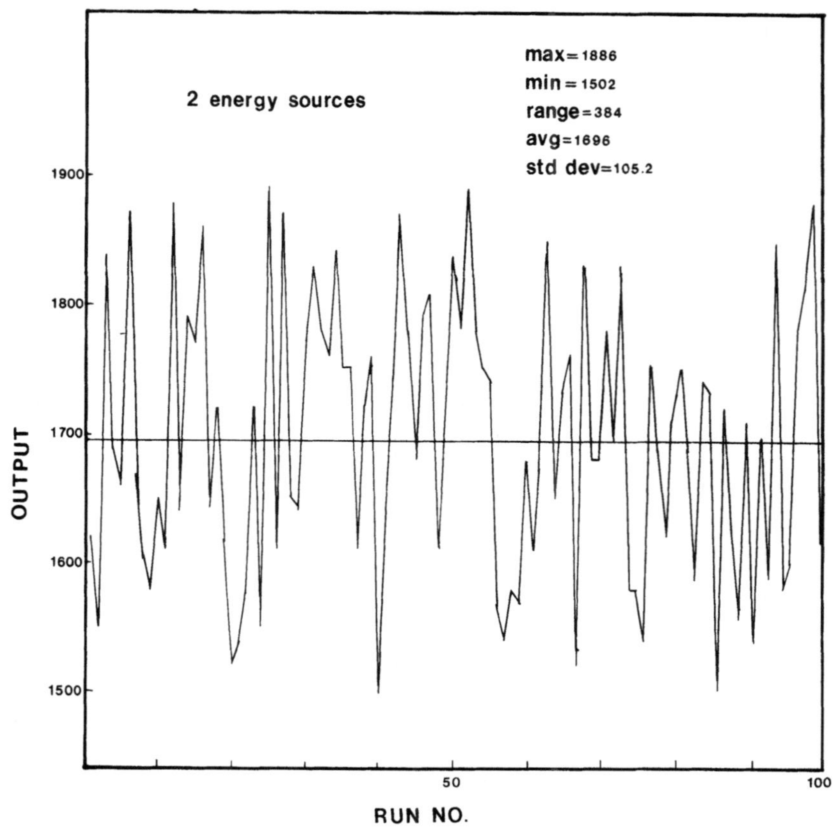

Figure 3.10. Total output of Gotland's economic system (in millions of Swedish crowns for 1975) for 100 simulation runs. Two energy sources and eight economic sectors with 25% reduction of interruption to all sectors except sector 4. Maximum probabilities of interruption were: fossil fuels, 30% and imported electricity, 15%.

able land, and levels of pollution to optimize the local supply of energy. It is assumed that these resources are only available to the extent that they are not used by the other sectors in the economic system; thus, there is competition only among the energy sources for the "left-over" local resources. The energy plantation possibilities included intensive and extensive operations on forest land or wetlands. The intensive cultivation implies both fertilization and irrigation. It was also assumed that the minimum amount of energy use should correspond to 1975 levels of consumption, which were considered to provide for an adequate standard of living. One model was chosen to explore the possibility of maximizing energy production from indigenous sources. It was formulated as follows (Ahlbom 1982):

$$\text{maximize } f = \Sigma\, a_j x_j.$$

x_j = energy supply from energy source j.

a_j = 1 if it is a local energy source; = 0, otherwise.

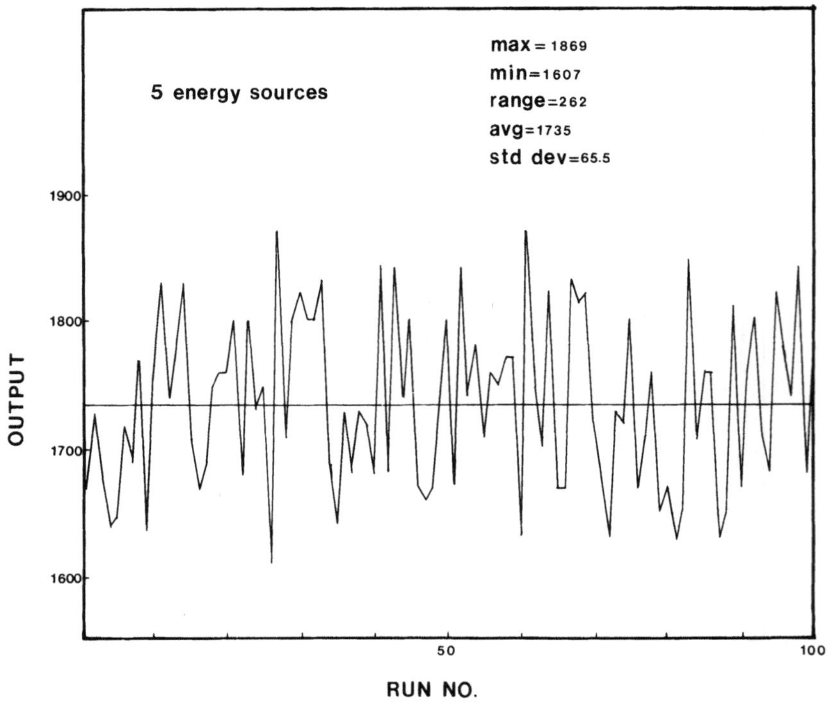

Figure 3.11. Total output of Gotland's economic system (in millions of Swedish crowns for 1975) for 100 simulation runs. Five energy sources and eight economic sectors with 25% reduction of interruption to all sectors except sector 4. Maximum probabilities of interruption were: (1) fossil fuels, 30%; (2) imported electricity, 15%; (3) wind-electric, 20%; (4) biomass, 0%; and (5) wave-electric, 20%.

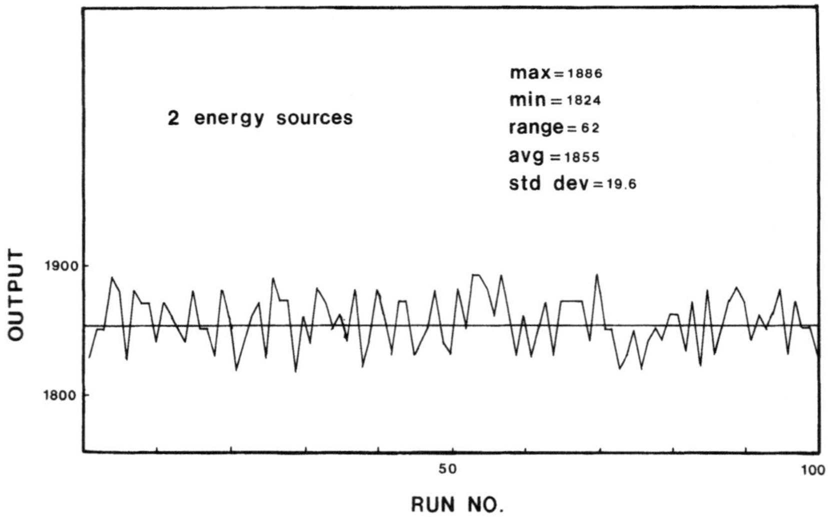

Figure 3.12. Total output of Gotland's economic system (in millions of Swedish crowns for 1975) for 100 simulation runs. Conditions are the same as in Figure 3.10 except that interruption to energy-efficient sectors is reduced to zero.

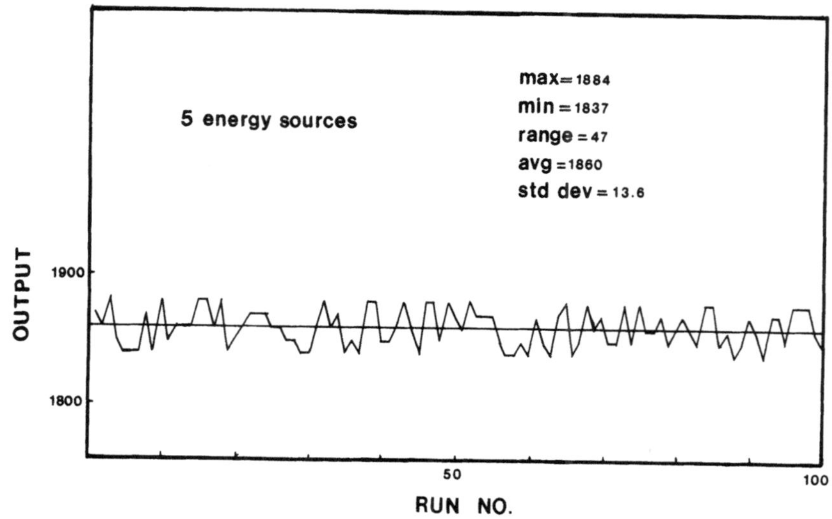

Figure 3.13. Total output of Gotland's economic system (in millions of Swedish crowns for 1975) for 100 simulations runs. Conditions are the same as in Figure 3.11 except that interruption to energy-efficient sectors is reduced to zero.

subject to:

1. Energy supply ≥ energy needed (1975):

$$\begin{aligned} \Sigma\, b_j x_j &\geq B \text{ (fuel)} \\ \Sigma\, d_j x_j &\geq D \text{ (gasoline and diesel)} \\ \Sigma\, e_j x_j &\geq E \text{ (electricity)} \end{aligned}$$

where b_j, d_j and e_j represent energy of each type per unit output from an energy source, j. For 1975, B = 7870 TJ, D = 1370 TJ, and D = 1720 TJ.

2. Resource constraints:

$$\begin{aligned} \Sigma\, t_j x_j &\leq \underline{T} \text{ (wetlands)} \\ \Sigma\, \bar{a}_j x_j &\leq A \text{ (agricultural land)} \\ \Sigma\, s_j x_j &\leq S \text{ (forest land)} \\ \Sigma\, k_j x_j &\leq K \text{ (land for energy plantations)} \\ \Sigma\, v_j x_j &\leq V \text{ (land for aerogenerators)} \\ \Sigma\, w_j x_j &\leq W \text{ (water)} \\ \Sigma\, z_j x_j &\leq Z \text{ } (SO_2) \\ \Sigma\, n_j x_j &\leq N \text{ } (NO_x) \\ \Sigma\, l_j x_j &\leq L \text{ } (CO_2) \\ \Sigma\, f_j x_j &\leq F \text{ (fertilizer)} \\ \Sigma\, m_j x_j &\leq M \text{ (imports)} \end{aligned}$$

Table 3.17. Coefficients for Optimization Model per GJ of Energy Output for Each Source (1980 Prices)

Energy Source	Water	Mire Land	Arable Land	Forests	Energy Plantation Land	Land for Wind-Electrical	SO_2	NO_x	CO_2
Energy plantations[a]									
IE	52	0	0	0	29	0	0	0	0
IM	52	29	0	0	0	0	0	0	0
EE	0	0	0	0	151	0	0	0	0
EM	0	151	0	0	0	0	0	0	0
Straw	0	0	846	0	0	0	0	0	0
Wood wastes	0	0	0	2800	0	0	0	0	0
Peat	0	11	0	0	0	0	100	440	89
Solar heating	0	0	0	0	0	0	0	0	0
Heat pumps	0	0	0	0	0	0	0	0	0
Oil	0	0	0	0	0	0	240	40	80
Coal	0	0	0	0	0	0	420	110	160
Wind-electric	0	0	0	0	0	60	0	0	0
Wave-electric	0	0	0	0	0	0	0	0	0
Electricity (cable)	0	0	0	0	0	0	0	0	0
Oil-electric	0	0	0	0	0	0	560	310	83
Coal-electric	0	0	0	0	0	0	400	310	106
Gasoline and diesel	0	0	0	0	0	0	0	0	0
Constraint	1.4×10^8	3.4×10^6	83×10^7	140×10^7	30×10^7	9×10^7	25×10^9	4×10^9	7.5×10^9
Units	0.001 m^3/day	m^2	m^2	m^2	m^2	m^2	gms	gms	kg

Energy Source	Capacities for				Need for			Fossil Fuel Import	Objective Function	
	Solar Heating	Heat Pumps	Wave-Electric	Electric (Cable)	Fuel	Gasoline	Electricity		Local Energy	Energy Cost
Energy plantations[a]										
IE	0	0	0	0	1	0	0	0	1	5.3
IM	0	0	0	0	1	0	0	0	1	5.3
EE	0	0	0	0	1	0	0	0	1	5
EM	0	0	0	0	1	0	0	0	1	5
Straw	0	0	0	0	1	0	0	0	1	3.9
Wood wastes	0	0	0	0	1	0	0	0	1	3.9
Peat	0	0	0	0	1	0	0	0	1	2.7
Solar heating	1	0	0	0	1	0	0	0	1	5.4
Heat pumps	0	1	0	0	1	0	0	0	1	4.4
Oil	0	0	0	0	1	0	0	0	0	4.9
Coal	0	0	0	0	1	0	0	0	0	2.5
Wind-electric	0	0	0	0	0	0	1	0	1	6.9
Wave-electric	0	0	1	0	0	0	1	0	1	8.3
Electricity (cable)	0	0	0	1	0	0	1	0	0	4.4
Oil-electric	0	0	0	0	0	0	1	3	0	5.7
Coal-electric	0	0	0	0	0	0	1	3	0	5
Gasoline and diesel	0	0	0	0	0	1	0	1	0	4.9
Constraint	70	270	145	3.99	7.87	1.72	1.37	7.19		
Units	TJ	TJ	TJ	PJ	PJ	PJ	PJ	PJ		öre/MJ

[a] IE = intensive energy plantations on forest land; IM = intensive energy plantations on wetlands; EE = extensive energy plantations on forest land; EM = extensive energy plantations on wetlands.

The smaller letters before the x_j's are per unit coefficients for each of the land categories, resources, and pollutants listed. The energy sources considered, coefficient values, and constraints are all summarized in Table 3.17. The amount of available water was determined by assuming a total irrigation capacity that included the construction of 10 dams. Estimates of biomass for energy supply from agriculture and forested areas were made with the assumption that existing levels of forest and agricultural output would not decline. Attention also was paid to impacts of fertilizer application on nitrate concentrations in the drinking water, which was regulated to not exceed 11.4 mg NO_3-N/liter. Based on the hydrologic modeling simulations (model 6) of fertilization and groundwater nitrate considerations, fertilization rates were set at 97 kg ha^{-1}, which is less than required for intensive energy plantations. Air pollution constraints were set at 10 times the total output of pollutants that existed in 1975. Oil imports were restricted to 75% of 1975 levels to simulate decreased availability.

Results were obtained for two base case optimization models, maximizing local energy production and minimizing total energy cost (Figure 3.14). For the case of minimizing total cost, the objective function was:

minimize $f = \Sigma\, c_j x_j$
where c_j = cost per unit of energy from energy source j (Table 3.18).

subject to the constraints listed above.

For the case of maximizing local production, the model chooses a large supply from energy plantations and a relatively small contribution from coal; electricity is completely supplied by wave- and wind-electrical sources. However, for the case of minimizing total economic costs, coal becomes the predominant fuel source and electricity is supplied by cable.

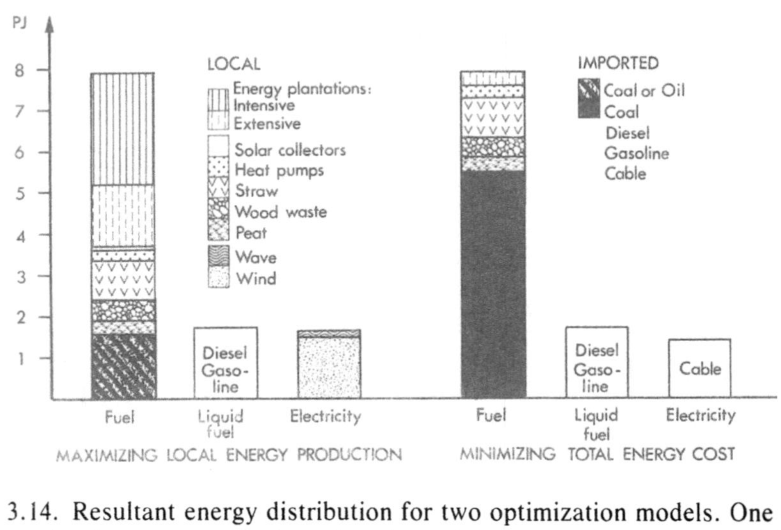

Figure 3.14. Resultant energy distribution for two optimization models. One model maximizes local energy production; the other minimizes total energy costs (Ahlbom 1982).

Table 3.18. Costs for Different Energy Sources in öre/MJ at 1980 Prices[a]

Source	Price (öre/MJ)
Peat[b]	2.7
Straw and wood waste[b]	3.9
Energy plantations[c,d]	
Extensive	5
Intensive	5.3
Solar heating[c]	5.6
Heat pumps[e]	4.4
Oil[b]	4.9
Coal[b]	2.5
Wind-electric[c]	6.9
Wave-electric[f]	8.3
Coal-electric[b]	4.7
Oil-electric[g]	5
Electricity from cable[e]	4.4
Gasoline[h]	4.9

[a] 100 öre = 1 Swedish crown
[b] Magnusson (1981).
[c] Anonymous (1979a).
[d] Lönnroth et al. (1977).
[e] Anonymous (1981a).
[f] Byström et al. (1980); Anonymous (1977a).
[g] Anonymous (1979b).
[h] Anonymous (1982).

Several simulations were performed to test the sensitivity of the model results to changes in critical parameters. With the objective of maximizing it, local energy production can supply all energy if demand is reduced by 25% or 50%. Higher levels of demand require increased imports of coal and oil (Figure 3.15). Another critical factor, as pointed out previously, is available water. Maximizing local energy production under conditions of different available water constraints during 4 months of the growing season illustrates the critical role that water plays for potential future energy systems (Figure 3.16). A doubling of water availability eliminates the need for imported fossil fuels, and a further expansion of available water would allow greater levels of total energy production than presently needed. Another critical factor, the level of fertilization, is seen to significantly affects the energy supply system through the effect on energy plantations; removing the constraints on fertilization greatly enhances the potential for local energy production (Figure 3.17).

Many other simulations were conducted with, perhaps, the most important point being that local energy production can supply demand, but at a cost as much as 50% higher than the strategy of minimizing total energy costs. One simulation maximizing supply from local energy sources under different values of a constraint on total energy cost showed that as total cost is allowed to

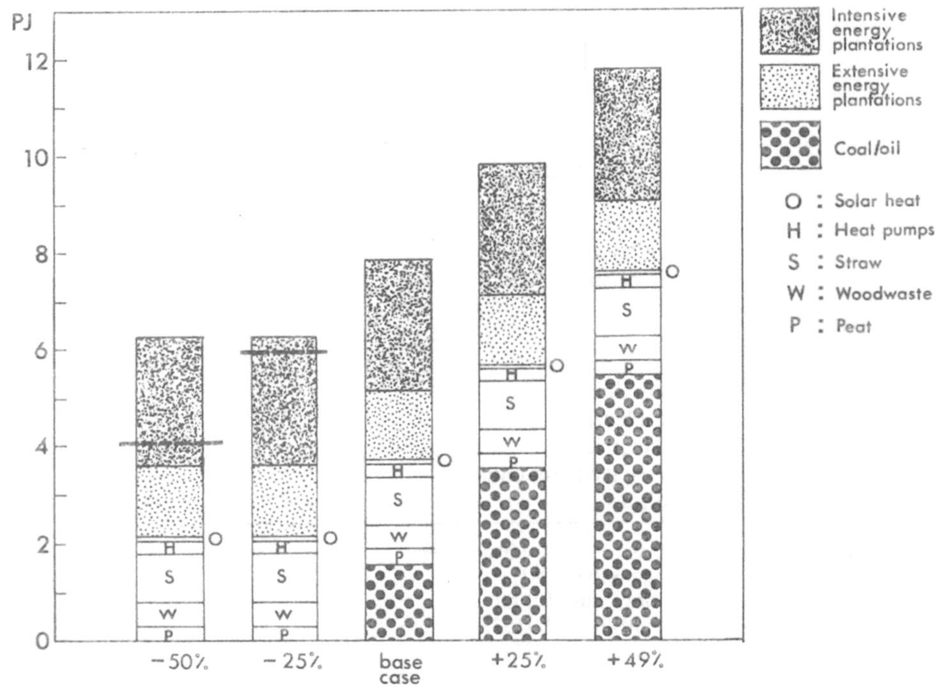

Figure 3.15. Results of the optimization model for maximizing local energy production with different levels of energy demand in comparison to the base case. For demand reduced by 25% or 50%, imports are eliminated; for demand increased by 25% or 49%, it is shown that coal and oil provide the supply. Amount above horizontal dashed line is surplus. (Ahlbom 1982).

increase, the energy system becomes less dominated by imported fuels and electricity and becomes increasingly dependent on local fuels and wind-electrical generation. The price of coal and oil is an important parameter in the models. Under conditions of minimizing total energy costs, as coal increases from the existing price of 2.5 öre MJ^{-1}, oil replaces it. As the combined price of coal and oil continues to rise, local energy sources become more dominant (Figure 3.18).

Model 6: Simulation Model for Assessing Water Quantity and Quality Impacts

As discussed previously, serious resource and environmental concerns on Gotland are related to the availability of sufficient water and the quality of groundwater, especially with regard to nitrate pollution. These areas of concern are coupled to energy and the economy in several ways. Enhanced energy availability, as well as demand for products from economic activities connected to agriculture, will stimulate production in various sectors. This will lead to increased demand for water and will result in impacts on the hydrology of the

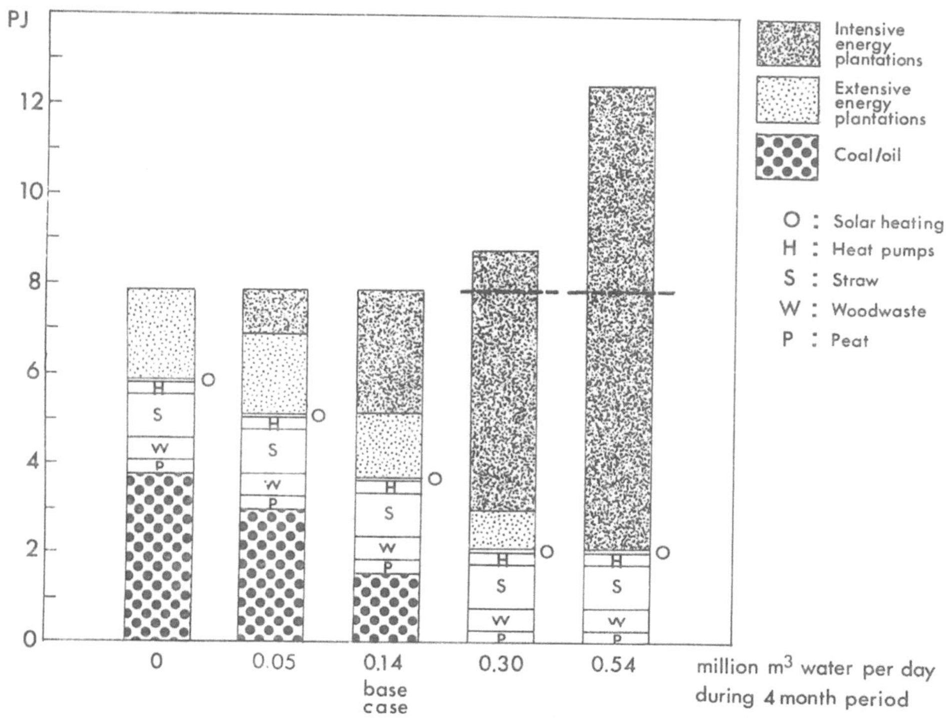

Figure 3.16. Results of maximizing local energy production under conditions of different levels of water available during a 4-month period. Amount above horizontal dashed line is surplus. (Ahlbom 1982).

island. If increased production in the agricultural system is pursued, then the use of more fertilizer, as well as irrigation water, will affect the island's water quality and quantity. Conversely, declines in economic activity due to declining exports might decrease demand for water and fertilizers with a positive impact on water quality and quantity. Modeling water quality and quantity is especially important for the enhanced economic activity scenario because of quality and quantity constraints on growth. A more direct link of water to the energy system arises if imported oil is to be replaced by indigenous energy sources, such as energy plantations. These will require extensive levels of irrigation and fertilization with consequent environmental impact. To analyze these relationships lumped parameter simulation models were formulated to assess the impact of different levels of irrigation and fertilization on the hydrologic cycle and groundwater quality (Jansson and Zucchetto 1978a, Spiller et al. 1981).

The basic model consists of storages of soil water, soil nitrate, soil ammonium, soil organic nitrogen, plant nitrogen, groundwater, and groundwater nitrate (Figure 3.19). These storages are affected by external forcing functions and interact with each other through various physical processes. Soil water is refilled through precipitation and irrigation, whereas it is lost through evapotranspiration, runoff, and percolation to groundwater. Fertilization increases

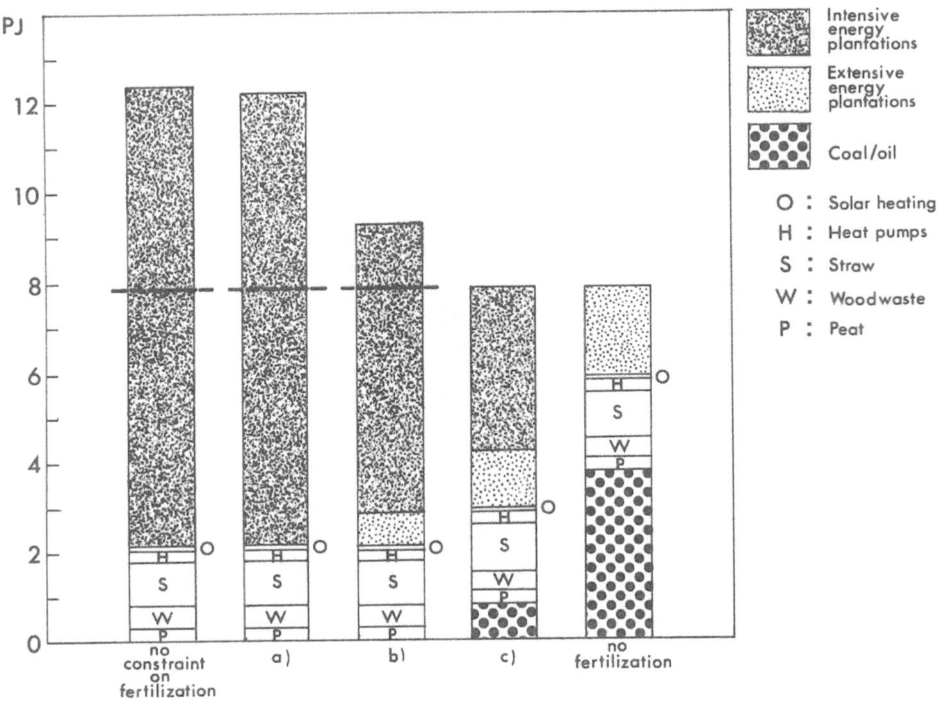

Figure 3.17. Results of maximizing local energy production under different constraints of nitrate concentration in groundwater. Cases a, b, and c represent 11.4, 8.8, and 6.8 mg/liter in groundwater, respectively. As this constraint is relaxed, local energy production is greatly enhanced. Amount above horizontal dashed line is surplus. (Ahlbom 1982).

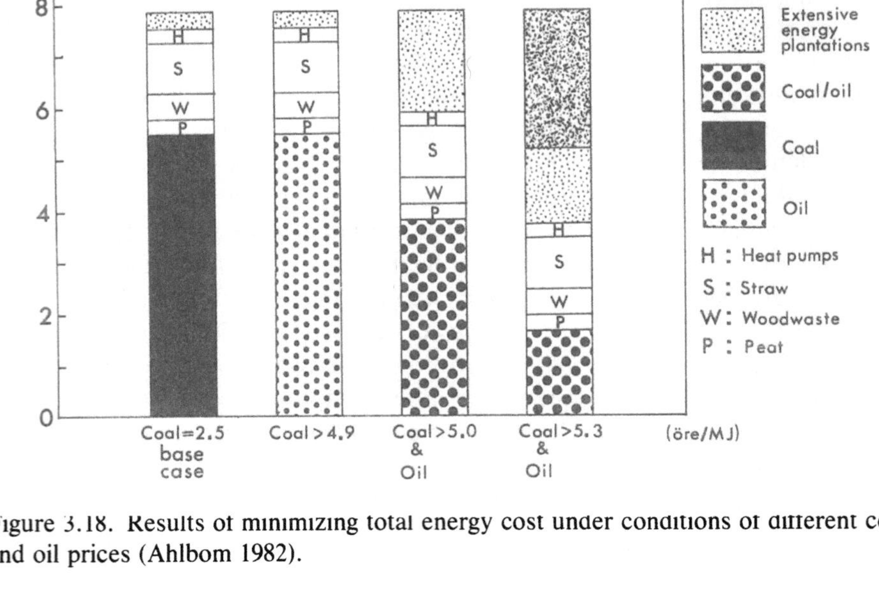

Figure 3.18. Results of minimizing total energy cost under conditions of different coal and oil prices (Ahlbom 1982).

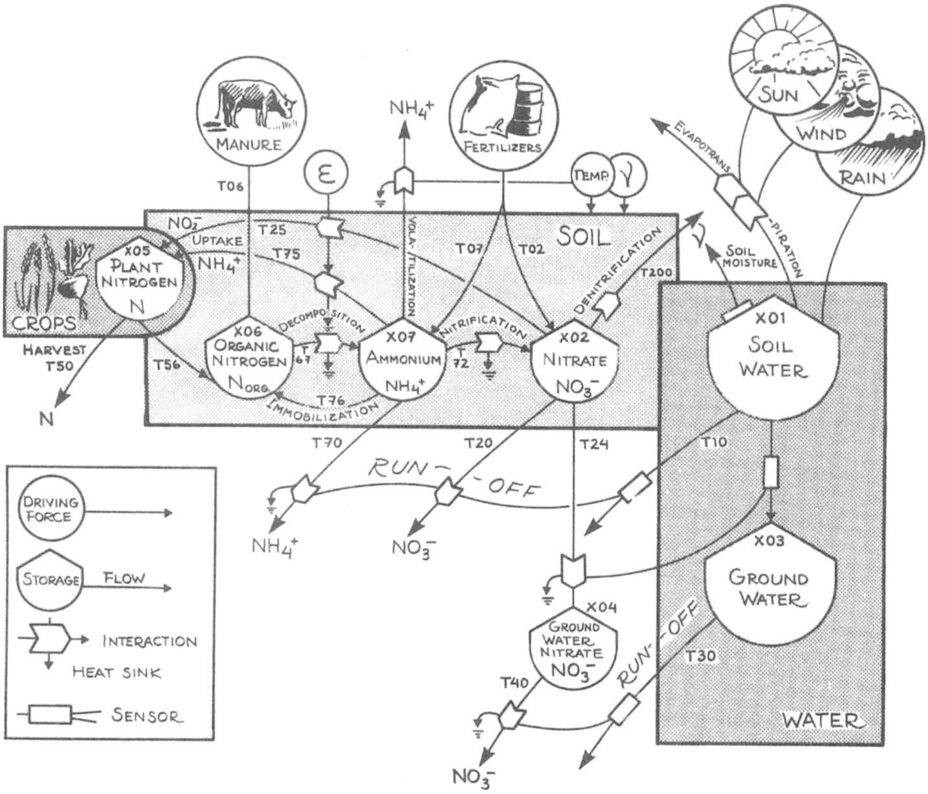

Figure 3.19. Water quantity and quality model for Gotland. See Table 3.19 for differential equations.

Table 3.19. Differential Equations for Water-Nitrogen Model in Figure 3.19

State Variable	Differential Equation
*XO*1 (Soil water)	$X\dot{O}1$ = (Precipitation) − (Evapotranspiration) − (Surface Runoff) − (Infiltration to Groundwater)
*XO*2 (Soil nitrate)	$X\dot{O}2$ = (Fertilizer) + (Manure) + (Nitrification) − (Plant Uptake) − (Infiltration) − (Denitrification)
*XO*3 (Groundwater)	$X\dot{O}3$ = (Infiltration to Groundwater) − (Runoff)
*XO*4 (Groundwater nitrate)	$X\dot{O}4$ = (Infiltration) − (Runoff)
*XO*5 (Plant nitrogen)	$X\dot{O}5$ = (Plant Uptake) − (Harvest) − (Crop Residue)
*XO*6 (Soil organic nitrogen)	$X\dot{O}6$ = (Manure) + (Crop Residue) − (Decomposition) + (Immobilization)
*XO*7 (Soil ammonium)	$X\dot{O}7$ = (Fertilizer) + (Manure) − (Nitrification) − (Immobilization) − (Uptake) − (Volatilization)

Table 3.20. Definition of Variables for the Differential Equations in Table 3.19

Associated Differential Equation State Variable	Variable Description
*XO*1	Precipitation: input of rain simulated on a daily basis (mm day^{-1})
	Evapotranspiration = $PET \bullet (S/SO)^{0.5}$
	where *PET* = potential evapotranspiration calculated from Penman (1948)
	S = water in top meter of soil minus the wilting point (mm).
	SO = field capacity-wilting point (mm)
	Surface runoff = $K \bullet$ Precipitation if (*XO*1 + Precipitation) > field capacity;
	K = 0.5 for soils with drainage;
	K = 0.1 for soils without tile drainage
	= 0 otherwise
	Infiltration to groundwater: occurs only in winter
	= Precipitation − XO1 if (*XO*1 + Precipitation) > field capacity
	= 0 otherwise
*XO*2	Fertilizer: rate of commerical fertilization (kg ha^{-1} day^{-1})
	Manure: rate of manure application (kg ha^{-1} day^{-1})
	Nitrification = $KN \bullet KNRF \bullet$ (XO7)
	where *KN* = 0.03 + 0.003 (T-20)
	T = temperature
	KNRF = 1 at 60% saturation, 0 at 0%
	Plant uptake = $UMAX \bullet \left(\frac{AET}{PET}\right)^{D(t)}$
	where *UMAX* = maximum nitrogen uptake for a given crop under optimal conditions (kg ha^{-1} day^{-1})
	AET = actual evapotranspiration
	PET = potential evapotranspiration
	D = 1 for first growth stage (8 weeks for barley), 0.5 afterwards.
	Infiltration = $\frac{XO2}{XO1} \bullet$ (infiltration to groundwater)

	Denitrification = $KD \cdot XO2$
	where $KD = KT_j \cdot KH$ if $XO1 >$ field capacity (soil-saturated)
	$KD = 0$ if $XO1 <$ field capacity
	where $KT_1 = 0.00125 + 0.000125$ (T-20)
	$KT_2 = 0.0006 + 0.00006$ (T-20)
	KT_1 = temperature coefficient for $T \geqslant 12°C$
	KT_2 = temperature coefficient for $T < 12°C$
	KH = pH coefficient = 1
$XO3$	Runoff = 0.07 • X03
$XO4$	Runoff (groundwater) = $K \cdot$ X04
	where K = 0.6 for high-porosity limestones
	K = 0.1 for low-porosity limestones
$XO5$	Harvest is just a function that removes plant nitrogen from system during time of harvest
	Crop residue is what's left after harvest
$XO6$	Decomposition = $KDE \cdot$ X06 (kg NH_4-N ha^{-1} day^{-1})
	where $KDE = KDMAX \cdot SMRF \cdot TRF$
	$KDMAX$ = optimal rate coefficient = 0.01 day^{-1}
	$SMRF$ = fractional reduction in decomposition rate at suboptimal soil water (1 at 60% saturation, 0 at 0%)
	TRF = fractional reduction in decomposition rate at suboptimal soil temperature (1 at 50°C, 0 at 0°C).
	If C/N of added organic material exceeds 25, no net decomposition occurs
	Immobilization occurs if $C/N > 25$. All ammonia ions produced are immobilized in microbial biomass
$XO7$	Nitrification = $KN \cdot KNRF \cdot$ X07
	where $KN = 0.03 + 0.003$ (T-20)
	$KNRF$ = soil reduction factor (1 at 60% saturation, 0 at 0%).
	Volatilization = $KTV \cdot KSM \cdot$ X07
	where KTV = temperature coefficient = 0.003 + 0.0003 (T-20)
	KSM = soil moisture coefficient (1 at 60% saturation, 0 at 0%)

soil nitrate and ammonium; soil nitrate is affected by nitrification of ammonium, denitrification, runoff, percolation, and uptake by plants. The nitrogen in plants is removed by harvesting or finds its way to the organic nitrogen storage, which is also enhanced by addition of manure, immobilization of ammonium, and decomposition. The model was formulated as a set of differential equations (Tables 3.19 and 3.20).

One important feature of this model is to capture interactions that occur in the hydrologic system due to human-induced changes in, for example, the forcing functions. One can simulate effects of agricultural practices on the island by changing rates of fertilization, irrigation, and groundwater withdrawal. This model can be used in several ways. Sensitivity analysis can be performed to

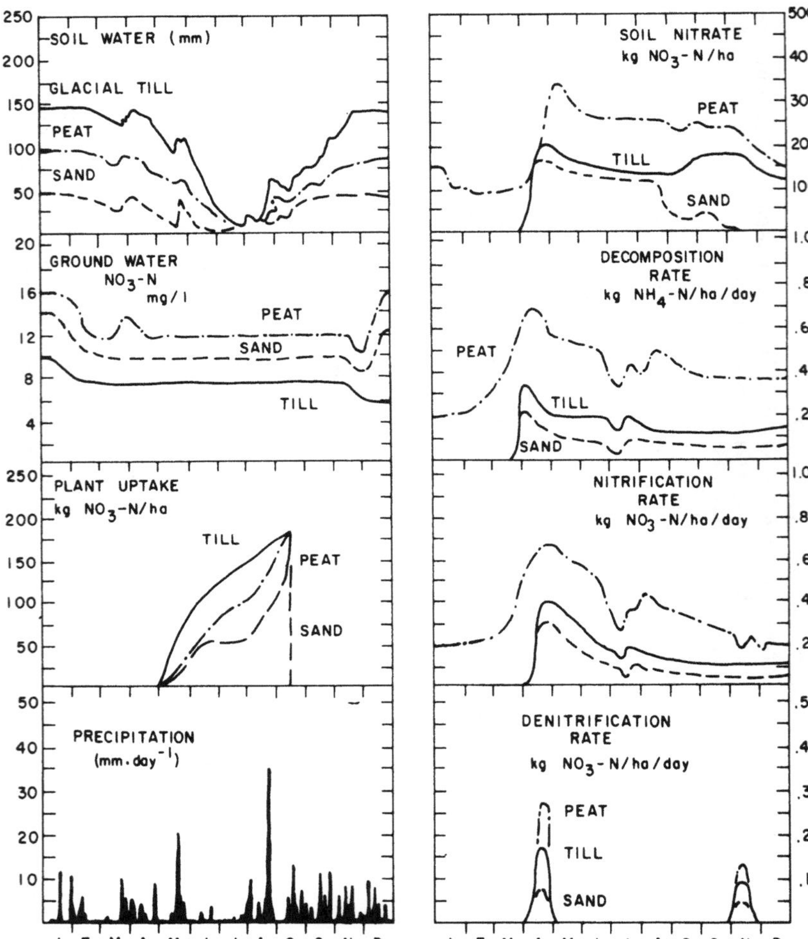

Figure 3.20. Simulation of barley growing on different soil types for 1972. Fertilizer applications (kg nitrogen ha^{-1} yr^{-1}) were: sand, 62; glacial till, 62 and peat, 0. Manure applications were 80 kg nitrogen ha^{-1} in spring and 20 kg nitrogen ha^{-1} in autumn (Spiller et al. 1981).

establish the relationship between levels of fertilization and groundwater nitrate concentrations. Other human activities affecting land use can also be simulated by changing rates of runoff and groundwater recharge.

Various kinds of regimes of irrigation and fertilization for different crops can be tested with this model to estimate, among other things, groundwater nitrate levels or soil water. Figure 3.20 presents simulation results for barley growing on different soil types, while Figure 3.21 shows the impact of changing the regime of fertilization and irrigation in comparison to standard practices. For sugar beets growing on glacial till, nitrate concentrations in groundwater increase linearly with fertilizer application (Figure 3.22). The coefficients relating agricultural output to nitrogen constraint in the economic and energy optimization models were derived from these simulation results.

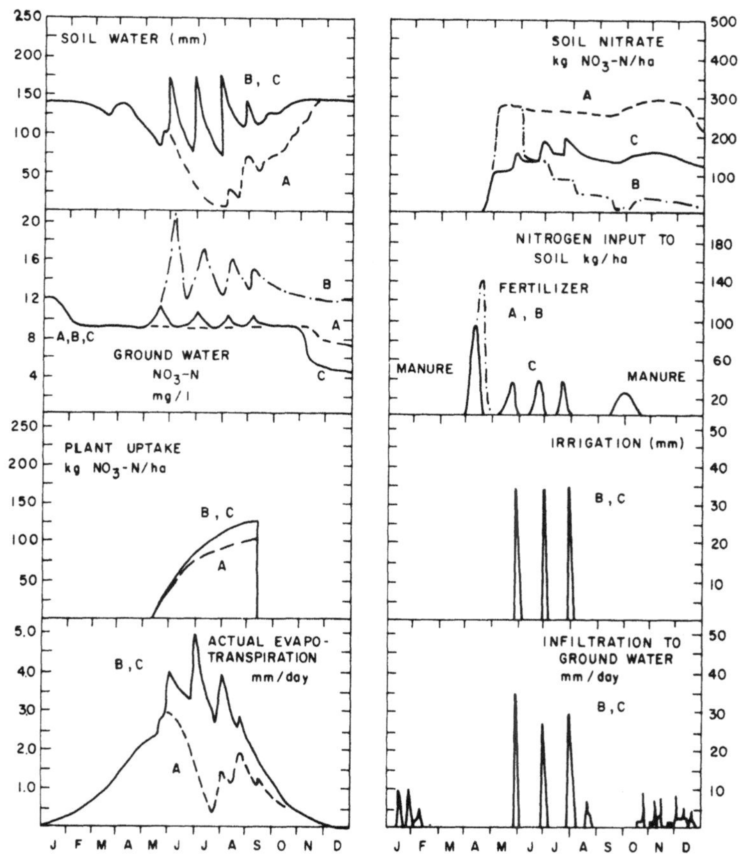

Figure 3.21. Simulation results for irrigation of sugar beets on glacial till. Three applications of 35 mm were made on June 1, July 1, and August 1. Curve A = standard run without irrigation; curve B = irrigation with a standard fertilizer application of 138 kg nitrogen ha^{-1}, curve C = irrigation with 90 kg nitrogen ha^{-1} of fertilizer distributed over three applications in May, June, and July. Manure applications were 80 kg nitrogen ha^{-1} in spring and 20 kg nitrogen ha^{-1} in autumn (Spiller et al. 1981).

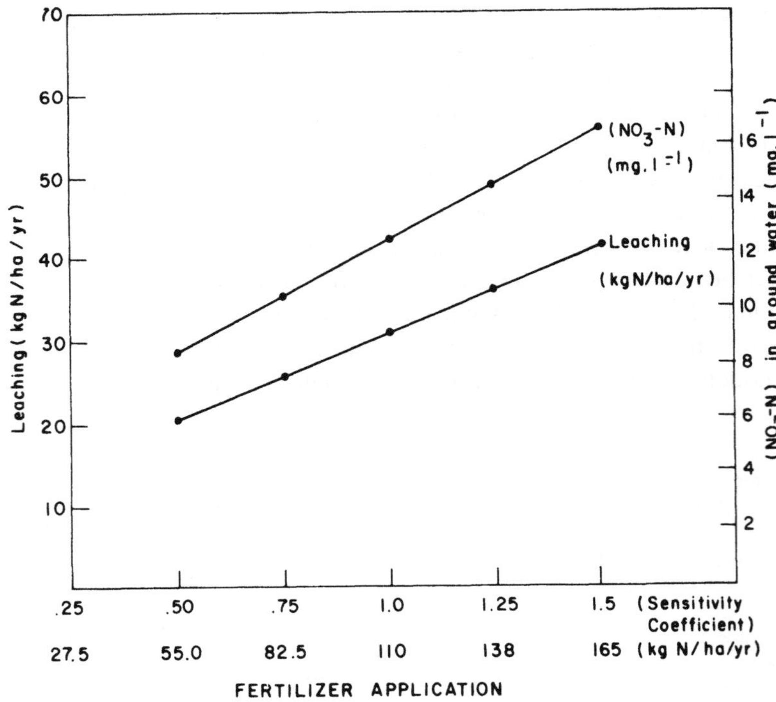

Figure 3.22. Sensitivity of predicted leaching and groundwater nitrate concentrations to changes in fertilizer application for sugar beets growing on glacial till (Spiller et al 1981).

In many ways, the hydrologic model is the most straightforward of all the models. Even though there is still a great deal of uncertainty about the limestone-based geology and its influence on the island's hydrology, the processes associated with water flow are well understood physically. This contrasts for example, with assumptions that have to be made in the optimization models. Since water is such a critical issue, we foresee that a spatial model of the island could eventually be formulated by dividing up Gotland into a number of subareas. A model for each subarea would be linked to its neighbors through flows of water. As briefly alluded to in the section in Chapter 2 on "The Study Site of Lummelunda," this hydrologic model was slightly modified and applied to the watershed of Lummelunda; therefore, it has applicability at different scales.

Model 7: Modeling the Coastal Ecosystem and Associated Human Impact

Our principal efforts dealing with ecosystem modeling were concentrated on the Baltic coastal system for a number of reasons. The Baltic Sea as a whole

Table 3.21. Inputs of Nitrogen to Coastal Waters Around Gotland[a]

Source	Input (Tons N/Yr)	Footnote
Atmospheric deposition	3800–41,200	b
Algal fixation	4490	c
Runoff	2850–3350	d
Up-welling	?	

[a] From Limburg (1983).

[b] Rohde et al. (1980) estimated inputs of nitrogen (N) from the atmosphere into the Baltic Sea as 120–1300 ktons/yr, or 0.328–3.55 tons/km²/yr. For Gotland's waters (area = 11,600 km², or 3.45% of the Baltic), the N input is therefore roughly 3800–41,200 tons/yr.

[c] For coasts, Lindahl et al (1978) estimated blue-green algal fixation at 0.6 tons/km² annually. Rinne et al. (1978) measured offshore fixation over 2 consecutive years, finding fixation in the summer of 1974 as 0.7 ton/km² and in the summer of 1975 as only 0.019 ton/km². Averaging these yields, 0.36 ton/km²/yr. Using Lindahl et al.'s estimate for the nearshore area (1300 km²) and the other estimate for the rest,

$$(0.6 \text{ ton/km}^2\text{/yr} \bullet 1300 \text{ km}^2) + (0.36 \bullet 10300) = 4480 \text{ tons/yr.}$$

[d] The lower estimate of runoff inputs from the island was done in the following manner. Long-term average monthly flow rates for four major drainage basins (SMHI 1979; SNV 1969), covering 21% of Gotland's surface area, were averaged to yield mean daily runoff rates. Next, monthly estimates of nitrogen concentrations in streams were made. Data for May, July, and December in 1973 and 1974 (Gotland County Administrative Office, Water Dept., unpublished) were used to calibrate an extrapolated annual picture of N concentrations, based on a likely seasonally-varying pattern. Finally, the average runoff data were combined with total N concentration estimates to produce an estimate of total nitrogen releases via runoff, as shown below:

Parameter	Jan.	Feb.	Mar.	Apr.	May	June	July	Aug.	Sept.	Oct.	Nov.	Dec.
Avg. runoff (m³/d/km²)	985	1033	938	1197	488	150	49	60	98	273	641	943
Runoff for Gotland (1000 m³/d)	3053	3203	2909	3710	1513	466	153	187	303	846	1987	2922
Estimated [N], mg/liter	5	5.25	5.25	6	2.0*	1.5	1.47*	1	0.25	0.5	2	5.24*
Total estimated N in runoff, ton/day	15.27	16.82	15.27	22.26	3.01	0.70	0.23	0.19	0.08	0.42	3.97	15.31

(*actual measured values)

The above yields an annual average daily nitrogen-loading rate of 7.8 tons/day, or 2847 tons/yr. That estimate may be low, due both to lack of sufficient data and to the fact that nitrogen-loading is likely to be greater now than in the early 1970s, since more fertilizer is being applied.

A higher estimate is found by using Nilsson's (1982) measurements of N in runoff in northwestern Gotland. His data yield an estimate of approximately 3350 tons/yr in runoff-loading.

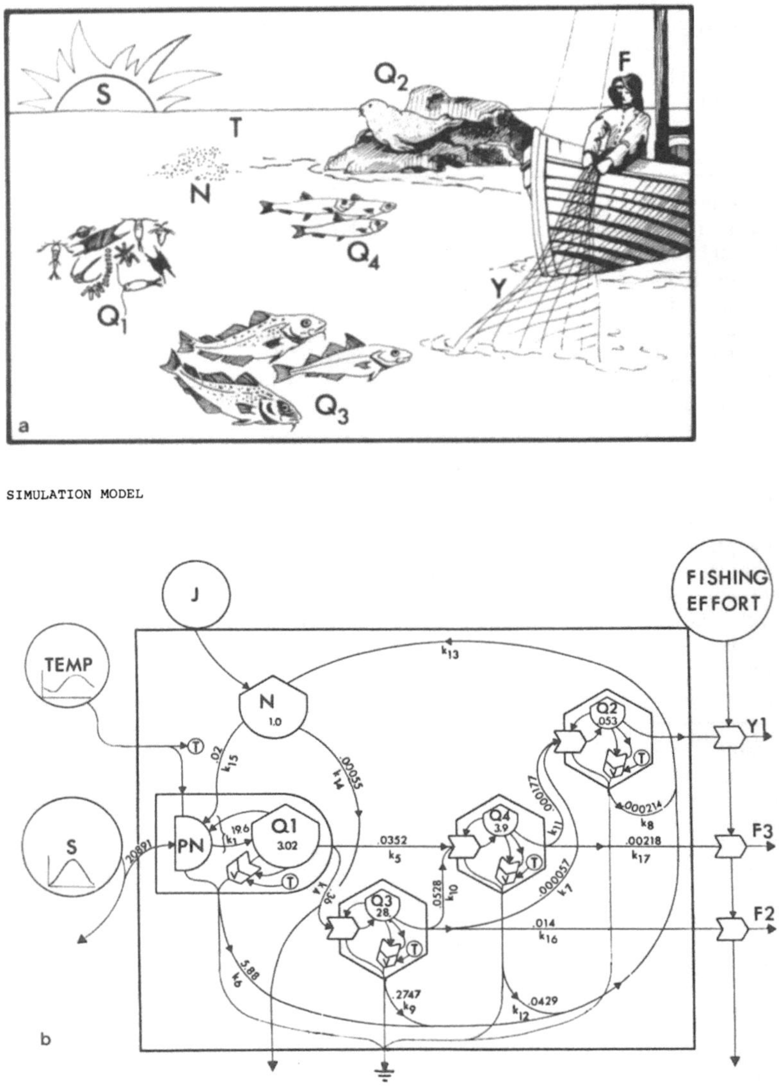

Figure 3.23. (a) Naturalistic description of the Baltic coastal ecosystem and (b) associated model diagram. Q_1 = primary producers; Q_2 = seals; Q_3 = herring; Q_4 = cod; J = input of nitrogen from terrestrial runoff; N = total nitrogen available for primary productivity; S = solar energy; T = temperature; F = fishing effort; and Y = harvest. Numbers refer to calculated coefficient values. See Table 3.22 for equations. Zooplankton, meio- and macro-fauna implicit in transfer from primary producers to fish. (Odum diagram adapted from Limburg et al. 1982).

Table 3.22. Differential Equations for Simulating the Gotland Coastal Ecosystem Minimodel[a]

(A) Forcing functions

Sunlight: $S = 10{,}666 - 10{,}236 \cos[2\pi(\text{day} + 9)/365]$ (kJ m^{-2} day^{-1})

- 10,666 = mean annual isolation (SMHI 1972) (kJ m^{-2} day^{-1})
- 10,236 = amplitude of fluctuation (kJ m^{-2} day^{-1})
- day = time interval with day = 0 on January 1
- 9 = number of days prior to January 1, which corresponds to minimum isolation (December 22)
- 365 = period of annual oscillation

Temperature: $T = 10 - 10 \cos[2\pi(\text{day} - 45)/365]$ (°C)

- 10 = both mean annual temperature and amplitude (°C)
- 45 = number of days after January 1 when the minimum temperature of 0°C is reached (February 14)

Van't Hoff equation (Sjöberg and Wilmot 1977)

$V = 2^{(T-10/K}, K \neq 0$

- K = increase in temperature required to double a process rate (here, K = 10°C)

Nitrogen runoff from land: $J = 0.0026$ (g N m^{-2} day^{-1})

(B) State variables

Primary producers (Q_1)

$dQ_1/dt = P_n - FOD1A - FOD1B -$ RESP1 (all flows in kJ m^{-2} day^{-1})

- P_n = photosynthesis = $P_{max}[S/(k_1 + S)][N/(k_2 + N)]Q_1 \cdot V$
- Michaelis-Menten formulation for light and nitrogen as limiting factors
- $FOD1A$ = herring (Q_3) grazing term = $k_4 Q_1 Q_3$
- $FOD1B$ = cod (Q_4) grazing term = $k_5 Q_1 Q_4$
- RESP1 = respiration = $kQ_1^2 V$

Herring (Q_3)

$dQ_3/dt = FOD1A - FOD2A - FOD3A - RESP3 - Y_2$

- $FOD1A = k_4 Q_1 Q_3$ = input from primary producers
- $FOD2A = k_7 Q_2 Q_3$ = portion preyed on by seals
- $FOD3A = k_{10} Q_4 Q_3$ = portion preyed on by codfish
- $RESP3 = k_9 Q_3^2 V$ = respiration/mortality
- $Y_2 = k_{16} Q_3 F_2$ = portion harvested by humans

Cod (Q_4)

$dQ_4/dt = FOD1B + FOD3A - FOD2B - RESP4 - Y_3$

- $FOD1B$, $FOD3A$ previously defined
- $FOD2B = k_{11} Q_2 Q_4$ = portion falling prey to seals
- $RESP4 = k_{12} Q_4^2 V$ = respiration/mortality
- $Y_3 = k_{17} Q_4 F_3$ = portion removed by fishery

Seals (Q_2)

The rate of change equation for seals is nearly identical to that for codfish, with two energy inputs and two outflows:

$dQ_2/dt = FOD2A + FOD2B - RESP2 - Y_1$

- $FOD2A$, $FOD2B$, $RESP2$ previously defined
- Y_1 = general term for a stock-depleting function, such as hunting or toxicologic stress (for the initial runs this was set to zero)

Nitrogen available for plant uptake (N)

$dN/dt = J + k_{13}\Sigma - SED - GRO$

- *where* $\Sigma = \alpha RESP1 + \beta RESP2 + \gamma RESP3 + \delta RESP4$ represents the sum of flows from primary producers, seals, herring, and cod, respectively; α, β, γ, δ are time-lag coefficients
- k_{13} = conversion factor from energy to gram nitrogen
- SED = sedimentation flow (assumed linear) = $k_4 N$
- GRO = nitrogen uptake by primary producers = $k_{15} P_n$

[a] Limburg et al. (1982).

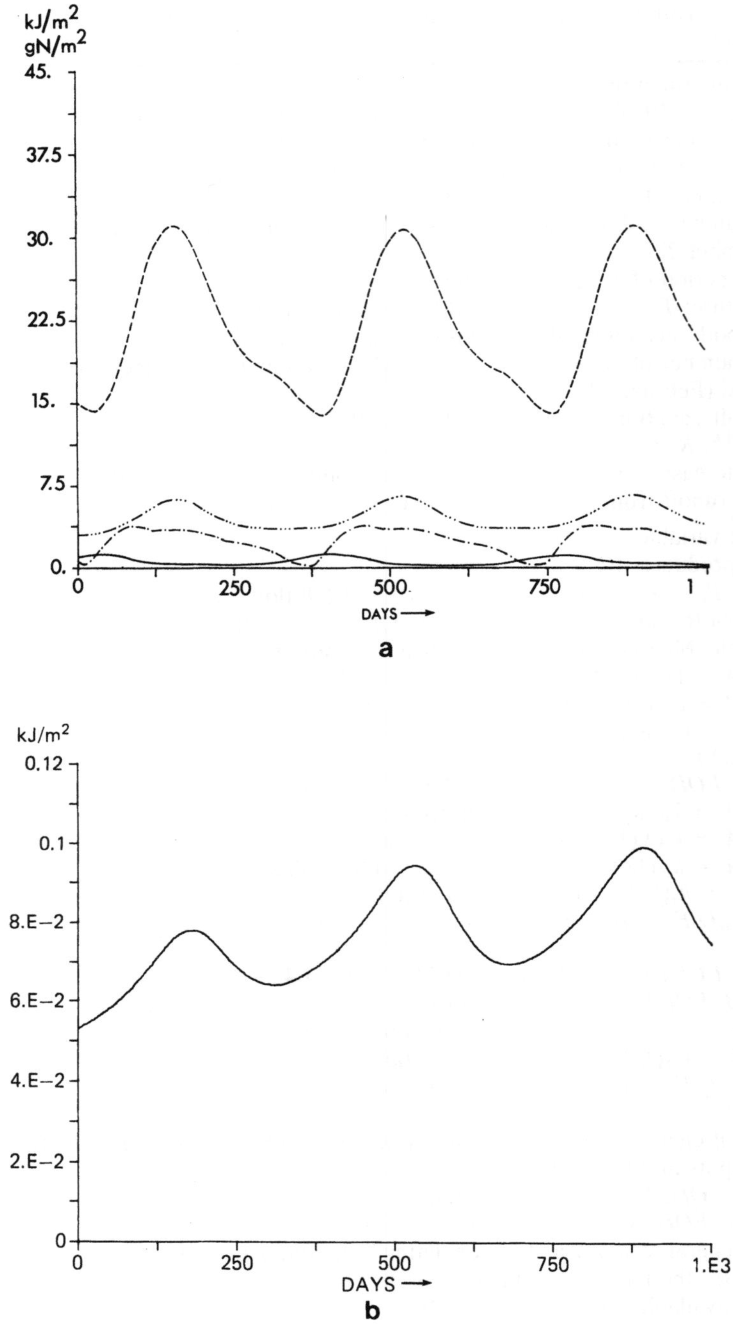

Figure 3.24. Results of the standard run for the model in Figure 3.23 over a 3-year simulation period. (a) Stocks of herring (----), cod (-····), primary producers (-·-·-), and nutrients (——). (b) Seal stocks.

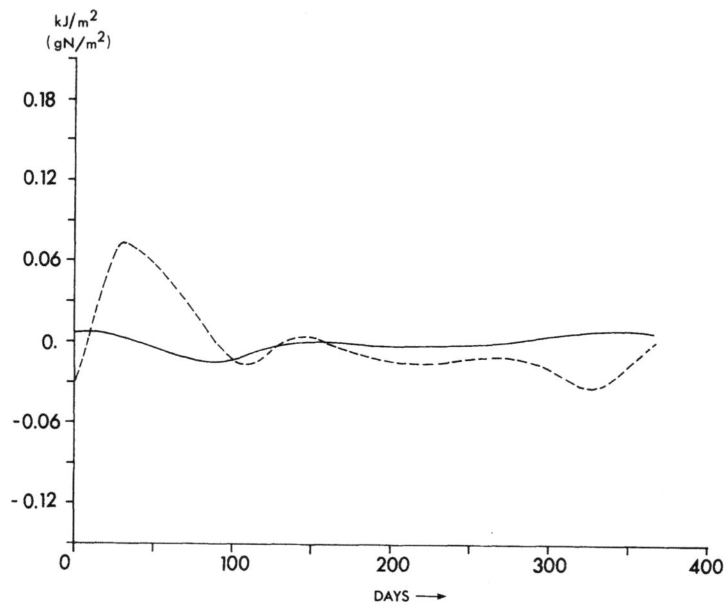

c

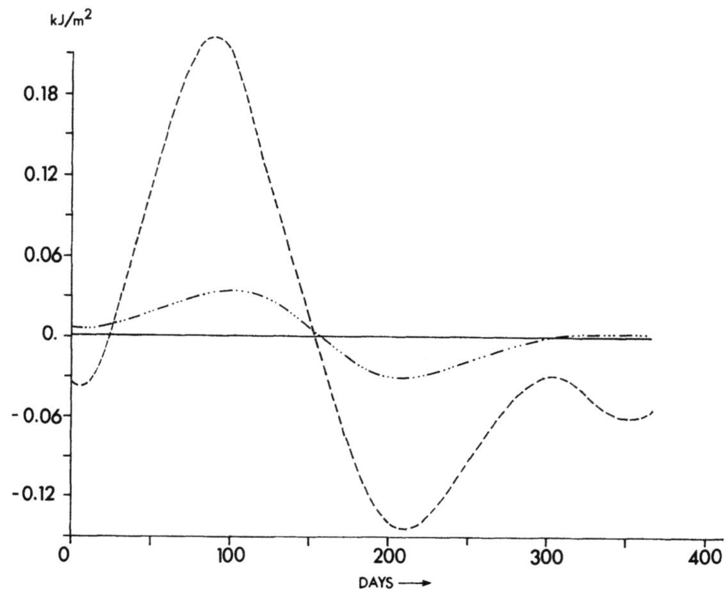

d

Figure 3.24. (c) Rates of change of primary producers (----) and nutrients(——). (d) Rates of change of herring (----) and cod (-····) (Limburg et al. 1982).

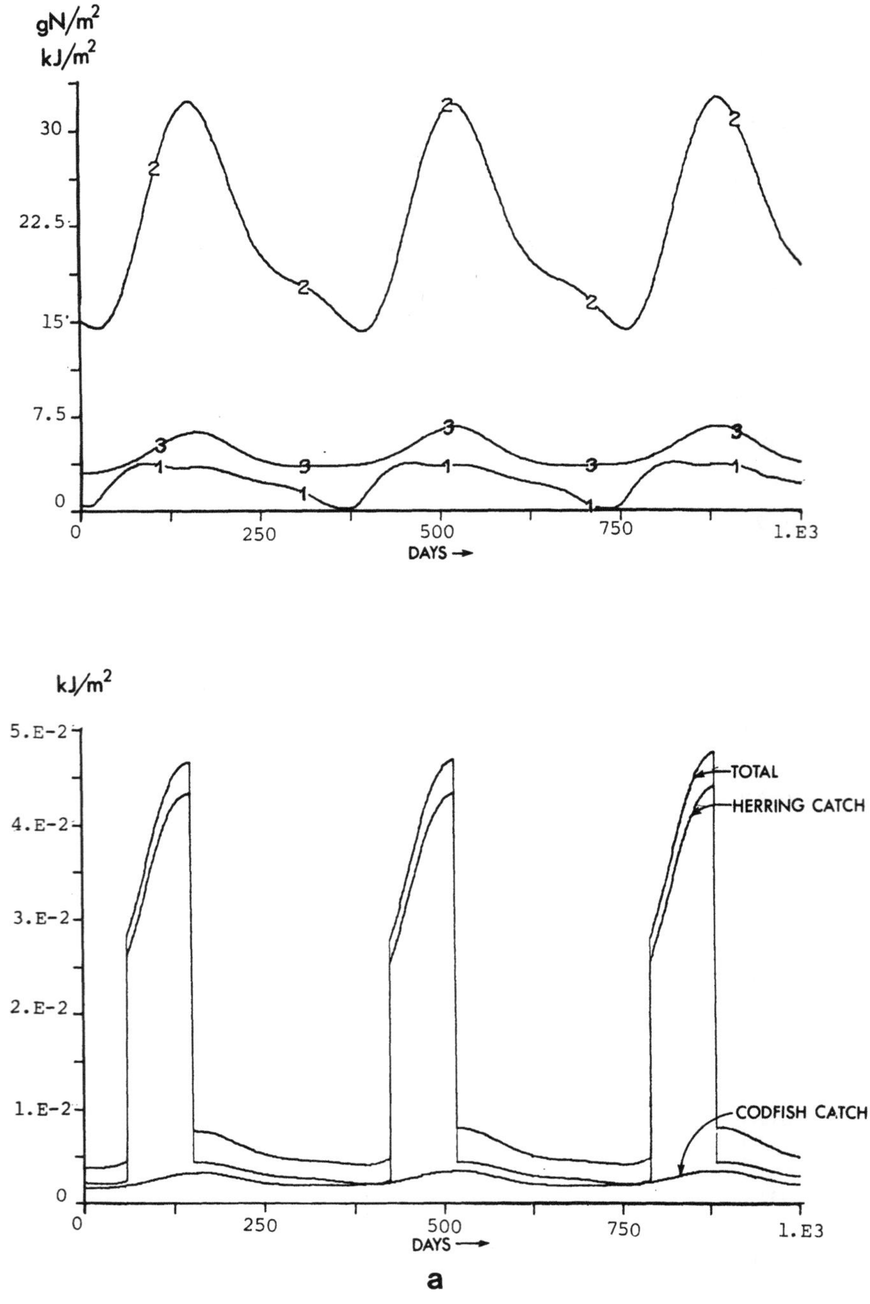

Figure 3.25. Simulation of model in Figure 3.23 with annual fishing effort concentrated over one-quarter of the year. (a) Maximum fishing effort during March-May with ecosystem response and fish yields.

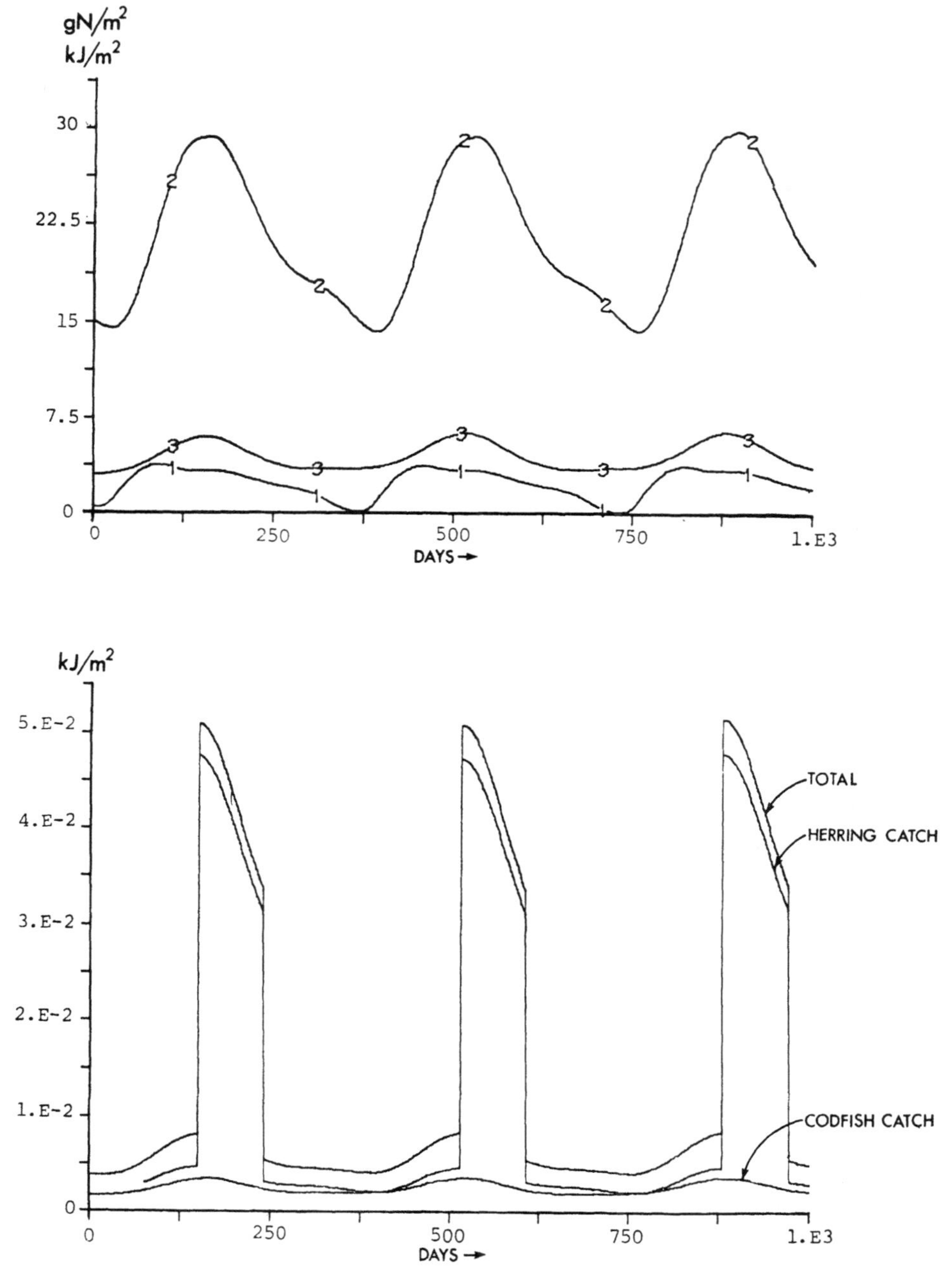

b

Figure 3.25. (b) Maximum fishing effort during June-August (Limburg et al. 1982). (1) primary producers; (2) herring; (3) cod.

Table 3.23. Sensitivity Analysis of Model for Changes in Nitrogen Input[a]

Run Description	J (g N m^{-2} day^{-1})	Annual Gross Production (kJ m^{-2} yr^{-1})	Average Annual P/R Ratio	Herring Catch (kJ m^{-2} yr^{-1})	Codfish Catch (kJ m^{-2} yr^{-1})
Standard run	0.0026	830	1.17	3.0	0.67
Run Ia[b]	0.026	2235	1.21	4.4	0.91
Run Ib[c]	0.052	3670	1.26	5.4	1.11
Run IIa[d]	0.136	890	1.17	3.0	0.68
Run IIb[e]	0.136	840	1.17	2.9	0.66
Run IIc[f]	0.136	770	1.16	2.9	0.66

[a] Limburg et al. (1982).
[b] Nitrogen flow J increased 10 times over standard run and distributed evenly over year.
[c] Nitrogen inflow J increased 20 times over standard run and distributed evenly over year.
[d] Annual inflow of nitrogen compressed into 1-week pulse, days 90–97 (April 1–7).
[e] Annual inflow of nitrogen compressed into 1-week pulse, days 180–187 (June 30–July 6).
[f] Annual inflow of nitrogen compressed into 1-week pulse, days 270–277 (September 27–October 3).

had been under study for many decades, with an unusual set of available basic knowledge and data (Jansson 1972). Furthermore, the extremely dynamic nature of that system and its historic importance for the support of fisheries and seal hunting amplified our interest in modeling the impact of human activity. There are several economic-ecologic interactions that are of interest. Naturally, the fishery forms a part of the economic system, and it influences the coastal ecosystem by changing the structure of the fish populations. The expansion of fish-processing industries on Gotland will have a set of economic consequences that could be estimated by using an input-output model for the regional economy. The expansion will lead to an increased demand for fish and to intensified fishing, with a resulting impact on coastal ecosystems. Other sources of human impact on the coastal ecosystem are sewage treatment and agriculture, both of which increase the amount of nutrient runoff from land. Estimates of nitrogen flows for the entire island were presented previously, but Limburg (1983) also made some further estimates for the coastal area within 12 nm around Gotland; these estimates were made on a monthly basis for the purposes of modeling the input of nitrogen to the ecosystem model (see Table 3.21). The total input of nitrogen from terrestrial sources may be as high as naturally occurring nitrogen fixation in the Baltic Sea, thus representing a significant flow.

The modeling of the coastal system proceeded in several stages with a sequence of models formulated (Limburg 1983). The first-generation models were formulated to capture the overall effects of nitrogen runoff and fish harvesting (Limburg et al. 1982). They included the following ecosystem components: primary producers, herring, cod, seals, and nitrogen storage. The flows modeled were inputs of solar energy and nitrogen and outflows due to human harvesting of herring, cod, and seals (Figure 3.23). A set of differential equations was used to represent this system (Table 3.22). Sunlight and temperature were indicated

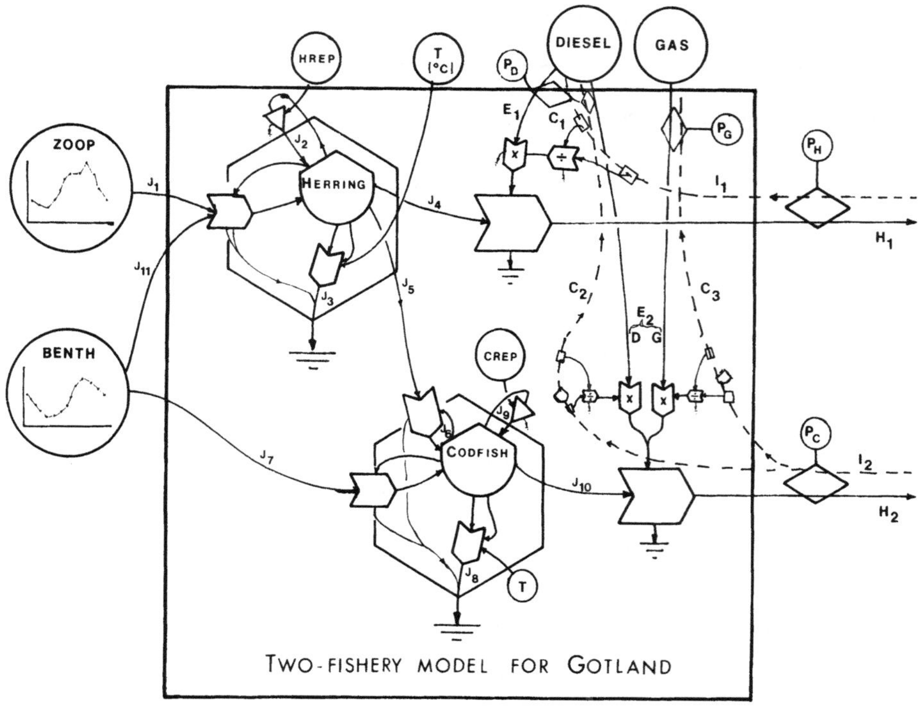

Figure 3.26. Model diagram for a simplified two-fishery model (TWOFISH). Flows are in g m^{-2} mo^{-1} for July 1975 unless otherwise stated (Limburg 1983).

with shifted cosine waves that represented seasonal variations. The effect of temperature was described using the van't Hoff (Q_{10}) equation with $Q_{10} = 2$. This temperature effect means that rates of activity double for every 10°C rise in temperature. Photosynthetic production was modeled by using a Monod-limiting factor pathway, with a Michaelis-Menten mathematic form. Respiration in all of the components included the effect of temperature, as well as a quadratic density-dependent flow. The nutrients (nitrogen) storage is modeled with inputs from land and fish and uptake via photosynthesis. Baltic herring (*Clupea harengus membras.*) accounted for about 59% of the 1973 catch by weight, whereas codfish (*Gadus morhua callarias*) comprised 18% (Jansson and Zucchetto 1978a). Although seals are not hunted now, they were included in the model for the purpose of assessing the impact on the coastal system of larger seal populations. The model was calibrated using the best data available in the literature (Limburg et al. 1982); therefore, for example, herring was found to account for 65% of food intake by cod, with the rest captured from lower trophic levels. Some results of the standard run for the model are presented in Figure 3.24.

Simulation of different fishing strategies indicated that harvesting herring at low biomass levels resulted in lower levels of both primary producers and cod

Table 3.24. Differential Equations for TWOFISH; A Model of the Cod and Herring Fisheries

Equations

Differential equation for herring biomass (H):

$$\frac{dH}{dt} = F1 + J2 - J3 - J4 - J5$$

Differential equation for cod biomass (COD):

$$\frac{d(COD)}{dt} = F2 + J8 - J9 - J10$$

Flows:

Herring feeding $= F1 = J1 + BF$

where $J1 = k_1 \bullet ZOOP_m \bullet H$ = zooplankton consumption

$BF = k_{6,m} \bullet BENTH_m \bullet H$ = benthos consumption

m refers to monthly values

$ZOOP$ = zooplankton biomass

$F1 \leq 0.5 \bullet H$, herring may only consume 0.5 times their weight/month.

Recruitment of juveniles $= J2 = k_2$ (see Table 3.28) (in June only)

Temperature-dependent respiration/excretion/mortality:

$$J3 = K_3 \bullet H \bullet V_1$$

$$V_1 = Q_{10,H}^{(T_m - 10)/10}$$

$$Q_{10,H} = A_1 \bullet T_m^{B_1}$$

T_m = monthly temperature

A_1, B_1 = empirically determined coefficients

$J4$ = herring from fishing mortality $= k_4 \bullet H \bullet E_1$.

$$E_1 = k_5 \bullet \text{diesel} \bullet (I_1 - C_1)$$
$$I_1 = P_H \bullet E_1$$
$$C_1 = P_D \bullet E_1$$
$$H_1 = k_{15} \bullet J4$$

where E_1 = amount of diesel fuel used in herring fishery

I_1 = income from catch

C_1 = cost of fuel

H_1 = total harvest

P_H = dockside monthly prices for herring

P_D = monthly prices for diesel

$J5 = k_7 \bullet H \bullet COD$ = predation by codfish

$F2 = J6 + J7$ = total consumption by codfish

$J6 = k_8 \bullet J5$ = predation on herring

$J7 = k_9 \bullet BENTH_m \bullet COD$ = predation on benthos

$F2 \leq 1.5 \bullet COD$ (cod stocks may only consume up to 1.5 times their weight per month)

$J8 = k_{10}$ = cod juvenile recruitment (in June only). See Table 3.28.

Table 3.24. *(Continued)*

$J9 = k_{11} \bullet COD \bullet V_2$ temperature-dependent respiration/excretion/mortality

$V_2 = Q_{10,c}^{(T_m - 10)/10}$

$Q_{10,c} = A_2 \bullet T_m^{B_2}$

A_2, B_2 = empirically determined coefficients

$J10 = k_{12} \bullet COD \bullet E_2$ = fishing mortality for cod

$E_2 = D + G$ = total energy used in fishery

$D = k_{13} \bullet \text{diesel} \bullet (I_2 - C_2)$

I_2 = income from catch

C_2 = diesel fuel cost

$G = k_{14} \bullet \text{gas} \bullet (I_2 - C_3)$

C_3 = gasoline cost

$I_2 = P_{cod} \bullet H_2$

P_{cod} = dockside monthly cod prices

$C_2 = P_{diesel} \bullet D$

P_{diesel} = monthly diesel prices

$C_3 = P_{gas} \bullet G$

P_{gas} = monthly gasoline prices

$H_2 = k_{15} \bullet J10$ = total codfish harvest

than if the herring population was allowed to grow to its summer peak before harvesting (Figure 3.25). Spreading out fish harvests over a 3-month period ameliorated the impact; system impacts depended very much on the timing of fishing effort. Because the existing seal population is so small, its removal caused no major change in the system. Increasing seal populations to 1000 (the estimated historic population size) results in both slightly reduced fish stocks and a cod consumption by the seals that is nearly twice the cod fishery yield of the mid-1970s. According to the model, increasing seal populations to 5000 is beyond the carrying capacity of the present coastal system. A sensitivity analysis, with respect to nitrogen input, was also conducted. It indicated, at least for this

Table 3.25. Initial Conditions for TWOFISH (see Table 3.24)

Parameter or Variable	Value
Herring (*H*)	3.5 (g/m^2)
Codfish (*COD*)	1.4 (g/m^2)
Diesel	20,000 (m^3)
Gas	1000 (m^3)
E_1	34 (m^3)
D	27.5 (m^3)
G	6.6 (m^3)

Table 3.26. Transfer Coefficients for TWOFISH

Coefficient	Value
k_1	0.010946
k_2	—[a]
k_3	0.94
k_4	2.87×10^{-4}
k_5	5.64×10^{-5}
k_6	—[b]
k_7	0.0154
k_8	0.95
k_9	0.0142
k_{10}	—[a]
k_{11}	0.80
k_{12}	0.0015
k_{13}	1.3×10^{-5}
k_{14}	5.42×10^{-5}
k_{15}	1350
Temperature coefficients:	
$A_1 = 1.9625$	
$A_2 = 1.3743$	
$B_1 = -0.387$	
$B_2 = -0.516$	

[a] See Table 3.28.
[b] See Table 3.27 for parameters with monthly values.

Table 3.27. Parameters with Monthly Values

Parameter	Zooplankton (g/m²)	Benthic Orgs. (g/m²)	k_6
Jan.	6.7	9	0.005
Feb.	2.7	8	0.005
March	5.4	9	0.005
April	5.9	10.5	0.001
May	4.9	15	0.001
June	20.9	20	0.001
July	22.0	23	0.01854
Aug.	21.2	24	0.01854
Sept.	33	22	0.01854
Oct.	15.4	19	0.008
Nov.	6.1	16	0.008
Dec.	7.1	11	0.008

Table 3.28. Recruitment Values for Herring and Cod (g/m^2 yr^{-1})

Year	Herring (k_2)	Cod (k_{10})
1972	1.42	0.20
1973	0.86	0.27
1974	0.92	0.32
1975	0.74	0.20
1976	0.44	0.20
1977	1.61	0.31
1978	0.51	0.54
1979	0.59	0.38
1980	0.43	0.16
1981	0.65	0.16

simple model, higher production and catches of herring and cod at increased nitrogen levels (Table 3.23).

Formulations of fishery models with more emphasis on economic aspects were also made (Limburg 1983), although with simpler ecosystem properties; single-species, two-species, and cohort models were considered. In this case, only the two-species model is presented and is referred to as TWOFISH. The model was a Lotka-Volterra-type formulation (Figure 3.26) to investigate such aspects as the effect of codfish as a predator on herring, the competition between herring and cod fisheries for economic resources, the allocation of resources to nearshore and offshore cod fisheries, and the effect of increased time resolution for the model. Interviews conducted with Gotland fishermen revealed that fish prices and fuel prices were the most important factors for fishing activity in the short run (monthly); thus, fuel usage was employed as a measure of fishing effort.

The ecosystem processes supporting the fisheries have been greatly simplified in this model. Population dynamics of the two species under consideration are emphasized and include feeding, metabolism, reproduction, predation, natural mortality, and harvest. The model consists of two differential equations; one for herring and the other for cod. (Table 3.24). Parameters were estimated with

Table 3.29. Simulated Results and Verification Data[a]

	Herring			Cod		
Year	Biomass	Catch (tons)	Real Catch (tons)	Biomass	Catch (tons)	Real Catch (tons)
1975	4.47	1246	1650	1.63	1135	965
1976	5.50	1558	1475	1.94	1331	1528
1977	5.51	1786	1930	1.91	1567	1308
1978	5.47	2506	2569	1.89	1662	1312
1979	5.67	2167	2682	1.99	1476	2316

[a] Biomass in g m^{-2}.

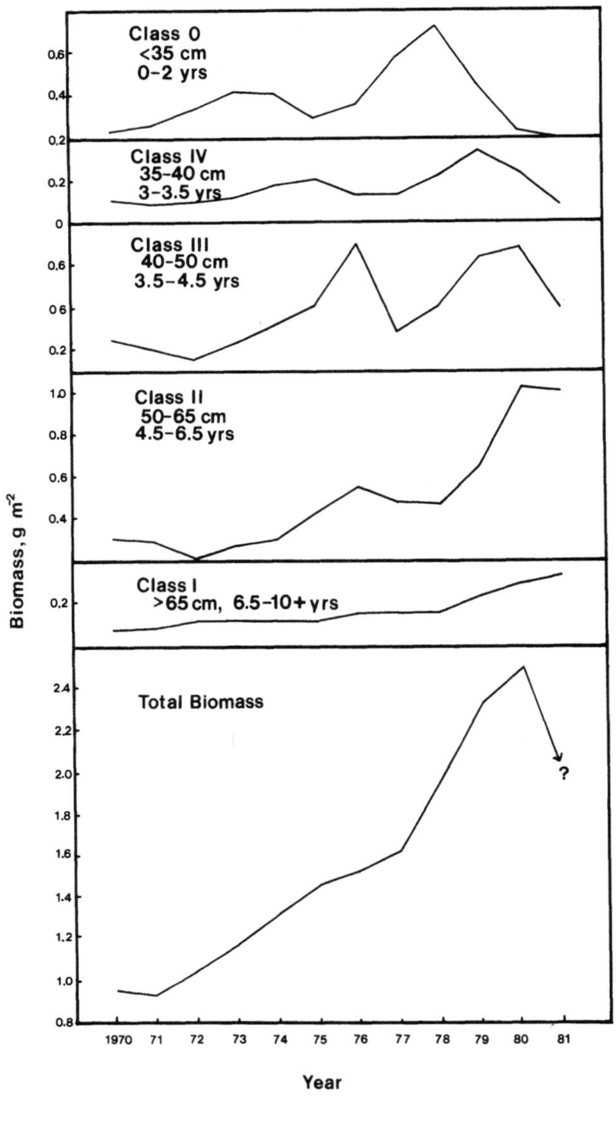

Figure 3.27. Estimated biomasses of codfish in the seas surrounding Gotland, as calculated by virtual population analysis. Note the movement of fish through different age cohorts, especially from 1978 (Limburg 1983).

real data or best estimates for July 1975. The forcing functions to the model were zooplankton and benthic faunal biomasses (representing the trophic web), fish recruitment, and prices for herring, cod, diesel fuel, and gasoline. These forcing functions were evaluated on a monthly basis and transfer coefficients were calculated (Tables 3.25–3.28). Results of one simulation are presented in Table 3.29.

The simulation model failed to predict the actual increase in codfish landings in 1979. In the real situation, large investments were made in the cod fishery after 1978 based on a good world market demand for codfish products as well as plans to locate a modern fish-processing plant on the island. The political strategy has been to maximize employment opportunities and job security in this traditionally insecure activity (Thiery 1979). In 1981, the processing plant was put into operation, and codfish harvests rose several hundred percent in comparison to 1979. In this model, fuel costs inhibited codfish catches to below observable levels. Actual codfish harvests continued to increase in the early 1980s due to unusually large Baltic cod populations; the model, as presently formulated, did not capture this trend (Figure 3.27). The model uncertainties include seasonal growth dynamics, migration behavior, and the effect of nutrient enrichment, as well as the dynamics of lower trophic level components such as zooplankton or the benthic fauna. These are all considerations that could be incorporated into more precise models, given that more adequate data eventually become available.

Plate 3. Drainage of wetlands has led to water quality problems, with increased nitrate values in groundwater near bogs (Photo, G. Håkansson).

4. Summary and Discussion of Results for Gotland

We have presented a great deal of data in Chapters 2 and 3 regarding our analysis of various activities on Gotland. These data have been organized in particular ways to emphasize resource, environmental, and economic measures. Readers of this volume may wish to interpret the data and model results in different ways. This is one reason for presenting extensive tables and graphs in the previous chapters and publications. However, in this chapter, we try to summarize the main results of the Gotland study and what they mean for the island and future planning. More general lessons regarding the study of large regions are postponed for later discussion in Chapter 5.

Energy and Development

Our historic analysis of development on Gotland confirms the notion that development seems to occur with transitions to higher quality and more abundant energy sources—a view that has been stated before in general terms by such investigators as Cottrell (1955) or Odum (1973). The renewable resources of fish, crops, wood, wind power, and water power provided and continue to provide sustenance to the local population. The relatively dispersed energy harvested in the Baltic Sea were superceded as a food resource for Gotland by the higher intensity activity of land-based agriculture. Note, in Figure 2.24, that photosynthetic production in agriculture has an intensity four times that of the

Baltic coastal systems. We have also seen agriculture develop towards more intensive use of the land. As discussed in the section in Chapter 2 on "Background and History," a shift in agriculture occurred around 200 A.D. from space-demanding exploitation to smaller farms and intensive land use. In more recent times, the drainage of wetlands of high photosynthetic intensities provided agriculture with fertile organic soils that enhanced agricultural production. Even more recently, fertilization and mechanization have increased per hectare yields to levels above those realized prior to World War II. Wood always provided fuel for heat, but when technologies allowed the eventual mechanical transformation of water flow and wind, additional energies became available. The relatively low gravitational potential energy of water sources was rapidly phased out as wind-powered devices provided greater yields, but also because of drainage activities that made small creeks dry up (see Figure 2.11). In the last century, the renewable energy sources were supplemented with coal, oil, and electricity—energy forms of higher quality and of greater abundance. Since World War II, the growing consumption of high-quality energy sources and the economy have rapidly transformed Gotland into a modern society. In particular, the cable from the mainland has made available a high-quality source of electricity that is diminishing the associated environmental impacts of electric power generation on Gotland. Is it no wonder, then, that local politicians have supported the decision to quadruple the capacity of this cable in a seemingly inexorable drive to capture high-quality sources of energy? This brief sketch of the energy development of the island seems to support the notion that systems tend to maximize useful energy inflows to grow, compete, and maintain complexity.

The consumption of energy on the island supports the production of economic value; maximizing this economic production corresponds to maximizing work done locally. Additionally, this helps to produce goods for export, generating income that can be exchanged for the import of goods. These imports required energy expended outside the borders of Gotland. Maintaining levels of economic activity near present conditions will require a continuing availability of high-quality energy. We have seen in Figure 2.18 that the energy-to-gross regional product (E/GRP) ratio was the same in 1979 as in 1972. This seems to imply that generating economic production at levels existing in the 1970s will mean similar levels of energy consumption. Improvements in energy efficiency can help to reduce to some extent the energy demand, but it will remain relatively high if any increased development is going to take place.

Trends in Energy Efficiency

The oil embargo of 1973 and the various energy price shocks of the 1970s have spurred efforts towards better uses of energy on Gotland as well as throughout Sweden. One measure of energy efficiency for economic systems is the amount of energy it takes to produce a unit of economic output; the less energy, the greater the energy efficiency. For the total economy of Gotland, the E/GRP ratio (see Figure 2.18) dropped sharply from 1973 to 1974. As of 1978–1979, this ratio reached levels equal to that in 1972 because of increasing electrical imports and their high value in oil equivalents. The decreases in this macroscopic indicator during the intervening years was partly due to higher energy prices precipitating a decrease in energy consumption at the household level. By some measures, this would imply a lower standard of living. Houses had to be insulated or kept cooler, people drove less, or generally constrained their activities. If attention is directed toward the industrial sector, trends in energy consumption to value added ratios (see Tables 2.20, 2.21, 2.23, and Figure 4.1) show that some industries have decreased in efficiency while others have increased; the overall trend from 1972 to 1979 has been one of increased energy efficiency. Much of this increased energy efficiency in the industry sector is due to technologic changes in the cement industry, which has reduced energy consumption significantly; they are also switching from oil to coal. In particular, the 42% decline in the ratio of fuels to value added for all industry sectors combined (see Table 2.26) contributed to a decreased dependence on imported fuels. This, coupled with an increased capacity of the mainland cable, amelio-

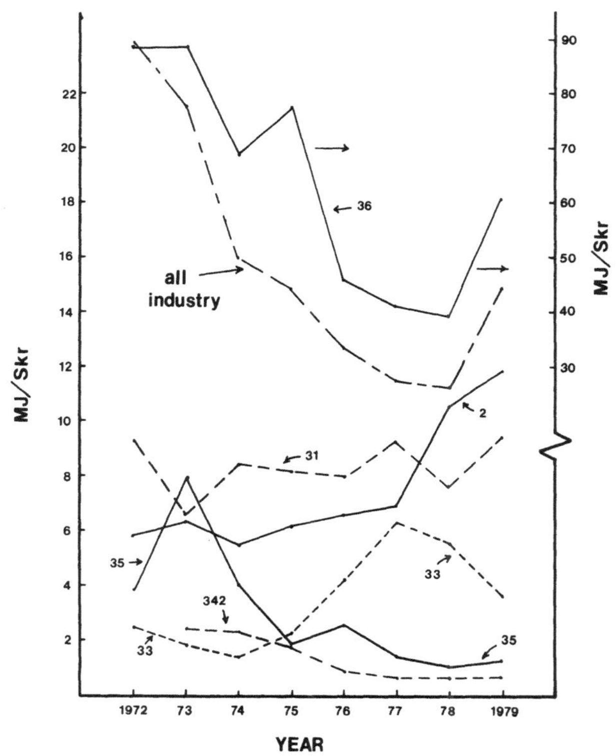

Figure 4.1. Trends in energy to value added ratios for selected industry sectors and for all industry sectors combined. Sectors are: 2, mining and quarries; 31, food; 33, wood; 342, graphics; 35, chemicals; 36, stone and soil. Units are MJ (oil equivalents) per Swedish crown at 1975 prices (see Table 2.25). The right-hand scale should be used for the stone and soil sector. For all other curves, use the left-hand scale.

rated the concern of interruptions in fuel imports. In addition, the planning authorities estimated in 1983 that conservation could save 16,700 m^3/yr of oil.

The renewable resource activities of agriculture and forestry should also be evaluated for trends in energy efficiency; however, sufficient time series data were not available to make any detailed judgments about trends in these activities. The analysis of fisheries (see Table 2.17) indicates a relatively stable level of direct energy cost per unit of catch from 1973 to 1977; this ratio dropped to 1.74 during 1978. For herring, the direct energy per unit of catch was less in 1974 than 1973, while being higher for cod (see Figures 2.26 and 2.27).

Energy-Economic Relationships

Much of the analysis presented in Chapter 2 was oriented towards measuring energy flows and economic activity in specific sectors. In this chapter, highlights of differences existing among different activities are addressed (Tables 4.1 and 4.2). In terms of direct energy consumption, we observe a range of input per unit of economic output, wages, and man-hours. The stone and soil sector is the most energy-intensive, while activities such as graphics or forestry are much lower on the energy-intensive scale. Most of the sectors have energy-to-value added ratios below 10 MJ Skr^{-1}, which is the average for the region (see Figure 2.18). Note the correspondence between the curves for industry as a whole and the energy-intensive stone and soil sector. Although Tables 4.1 and 4.2 are not strictly comparable, there is a definite decline over time in energy-to-value added ratios for stone and soil and the chemical sectors; sectors such as mining

Table 4.1. Comparison of Energy-Economic Ratios for Different Activities in 1972[a]

Activity	DE/V_A (MJ Skr^{-1})	$(DE + IE)/V_s$ (MJ Skr^{-1})	DE/V_w (MJ Skr^{-1})	DE/labor (GJ man-hr^{-1})
Regional economy	13.2	—	—	—
Stone and soil	101.7	58	370.7	11.3
Food	8.69	1.97	16.3	0.44
Quarries	8.19	5.21	15.6	0.5[b]
Wood	2.83	1.82	3.4	0.07
Textiles	1.78	0.69	1.96	0.06
Workshops	1.33	3.1	2.22	0.05
Agriculture	2.7	4.01	7.38	0.06
Forestry	1.18	—	1.63	0.04
Fisheries[c]	13.38	11.77	13.8	0.11[d]

[a] DE = direct energy (oil equivalents); IE = indirect energy or energy embodied in imported goods (oil equivalents); V_A = value added; V_s = value of sales; and V_w = value of wages.
[b] Data for 1973.
[c] Energy data for 1973; economic data for 1972.
[d] Value for 1973. Full-time fishermen work 40 hr/wk; part-time fishermen, 10 hr/wk.

Table 4.2. Comparison of Energy-Economic Ratios for Different Activities, 1977[a]

Activity	DE/V_A (MJ Skr^{-1})	DE/V_w (MJ Skr^{-1})
Regional economy	9.48	—
Stone and soil	40.98	202[b]
Food	9.15	20.35
Mining and quarries	7.98	28.5
Wood	6.25	9.23
Textiles	2.38	3.6
Chemicals	1.32	3.17
Workshops	1.09	7.07[b]
Graphics	0.68	1.85

[a] Economic data evaluated at 1975 prices; DE = direct energy (oil equivalents); V_A = value added; V_w = value of wages.
[b] Data for 1978.

and quarries, wood, textiles, and workshops seem to have declining ratios (see Table 2.25). From the point of view of wages, jobs, and energy, it is evident that equivalent levels of energy consumption generate widely different levels of wages or man-hours. We see from Tables 4.1 and 4.2, for example, that direct energy consumption in fisheries generates more man-hours and wages than in sectors such as quarries, food, or stone and soil. A somewhat different breakdown by activity was included in Chapter 3, model 1, which dealt with the calculation of resource requirements.

If both the indirect and direct energy requirement's per unit of output (this is different from value added considered above) from each sector are considered, then it can be seen that the electricity and water sector has the highest energy requirement (see Table 3.5). Indirect energy consumption is also important for many other sectors such as 1.1 (animal), 2 (mining), 3111 (slaughtery), 3112 (dairy), 3.1 (other food), 3.3 (wood), and 3.4.2 (printing). If the total energy requirement (direct and indirect) in both fuels and electricity (in oil equivalents) is calculated along with total employment and value added per unit of output from each sector, this information allows the construction of Table 4.3. This table indicates the energy-intensive sectors (e.g., 4, 3.6.9, 2, and 3118) and those generating high employment. It is interesting to note that the value added per unit of output is reasonably uniform across sectors (average equals 0.73; mean deviation equals 0.115), so it is equally attractive to expand any of the sectors using this measure as a criterion. Energy-to-economic ratios can also be formed from these data to assess relative energy-to-labor requirements and value added generated per unit of output from each sector (Table 4.4). Such sectors as forestry, construction, and services require much labor in relation to energy consumed. These sectors also require relatively small amounts of energy per unit of value added, as does the communication equipment sector.

Table 4.3. Total Resource Requirements per Swedish Crown of Output, 1975

	Sector	Energy[a] (MJ Skr^{-1})	Employment[b] per 1000 Skr	Value Added[b] per Skr
1.1	Animal	3.93	0.025	0.81
	Crop	3.99	0.024	0.58
1.2	Forestry	1.2	0.014	0.74
1.3	Fishery	7.97	0.025	0.52
2	Mining	10.04	0.009	0.91
3111	Slaughtery	3.74	0.021	0.82
3112	Dairy	6.14	0.022	0.83
3118	Sugar	10.7	0.018	0.71
3.1	Other food	4.14	0.015	0.61
3.3	Wood	3.6	0.021	0.87
3.4.2	Printing	2.79	0.013	0.92
3.6.9	Stone and soil	46.4	0.005	0.58
3.8.3	Communication equipment	0.97	0.005	0.80
	Other workshops	2.79	0.016	0.75
4	Electricity and water	68.01	0.014	0.35
5	Construction	1.28	0.015	0.83
61/62	Wholesale	3.2	0.024	0.89
7	Transports	4.24	0.015	0.71
	Tourism	9.23	0.019	0.66
8/9	Services	1.77	0.019	0.70

[a] In oil equivalents. See Table 3.5 for total oil and total electricity. Energy = total oil and total electricity ÷ 0.3.
[b] From Table 3.5.

Table 4.4. Ratios of Energy to Employees and Energy Required to Value Added Generated, 1975[a]

	Sector	Energy per Person Employed (kJ Person^{-1})	Energy/ Value Added (MJ Skr^{-1})
1.1	Animal	157.2	4.85
	Crops	166.3	6.88
1.2	Forestry	85.7	1.62
1.3	Fishery	318.8	15.3
2	Mining	1115.6	11
3111	Slaughtery	178.1	4.56
3112	Dairy	279.1	7.4
3118	Sugar	594.4	15.07
3.1	Other food	276	6.79
3.3	Wood	171.4	4.14
3.4.2	Printing	214.6	3.03
3.6.9	Stone and soil	9280	80
3.8.3	Communication equipment	194	1.21
	Other workshops	174.4	3.72
4	Electricity and water	4857.9	194.3
5	Construction	85.3	1.54
61/62	Wholesale	133.3	3.6
7	Transports	282.7	5.97
	Tourism	485.8	13.98
8/9	Services	93.2	2.53

[a] Constructed from Table 4.3 by dividing energy column by employment and value added columns, respectively. Energy is in oil equivalents.

Economic Stability and Dependence on External Sources of Energy

As discussed in Chapter 2, Gotland is almost totally dependent on imports of fuels and electricity for support of the economic system. Emerging trends or possible modifications of existing arrangements can lead to a reduction of this dependency and to a decreased susceptibility to imported energy interruptions or price increases. One encouraging trend is the increase in energy efficiency that has occurred for some of the major industrial sectors. During the energy crisis of 1974, and on until 1977–1978, a sharp reduction in imported oil occurred (see Table 2.3). However, we see a significant rise into 1981 of liquid fuels, as well as an increasing dependence on coal. Imported fuels in 1972 amounted to 13,828 TJ and to 14,692 TJ in 1981—an actual overall increase in comparison to the decline that occurred in the mid-1970s. Electricity imports also increased from 555 TJ in 1972 to 846 TJ in 1980. The E/GRP ratio decreased immediately after the 1974 energy crisis, but it reached preembargo levels in 1978 and 1979 (see Table 2.8). In the industry sectors, fuel consumption was about 10% lower in 1979 than in 1972; however, electrical consumption was 33% higher (see Table 2.24). These data indicate that although there has been some improvement in energy efficiency in the industrial sector, the economy is still heavily dependent on imported fuels. We have not analyzed the economic data of 1979–1981 to compare changes in energy-to-economic measures for these years; thus, we have not assessed the energy efficiencies in the 1980s. In recent years (1978–1981) imports of coal have replaced part of the oil used in cement production (see Table 2.3). The planning authorities envision several ways that import dependence on oil will be reduced. These are increasing imported electricity via the mainland cable, cement production using coal instead of oil, coal-fired district heating in Visby, conservation, electrical heating in small homes, heat pumps, district heating using wood chips and waste heat, and local oil production.

Several of the presented models explore the impact of energy interruptions, diversification, and the contribution of potential endogenous renewable energy sources (see models 2, 4, and 5 in Chapter 3). The regional optimization (model 2) for a simplified Gotland economy describes postulated scenarios of price increases, oil interruptions, and energy plantations over a 20-year period (Table 3.10). Under the objective of maximizing value added, oil disruptions severely impact on such sectors as stone and soil, quarries, and electric power generation, while agriculture, food industries, workshops, and wood industries (for example) maintain or increase their levels of activity. Furthermore, as oil prices rise, local energy plantations become economically competitive; although they produce a small percentage of the total energy requirements, employment is greatly enhanced (see Figures 3.4 and 3.5).

The partial transition to coal seems to be a beneficial development not only with regard to energy prices, but also concerning energy availability. The energy diversity models (model 4) showed to what extent diversifying across energy supply sources and end-use activities can diminish fluctuations of total economic output under conditions of uncertainty in energy supply, as well as raise the average value of economic production. Deliberate policies of allocating available energy to energy-efficient sectors can substantially improve the stability of economic output. Thus, both diversification of energy sources and allocation schemes were concluded to be effective strategies for improving economic performance. However, this diversification engenders a greater cost. Based on the energy prices in Table 3.18 and the energy supply structures in Table 3.17, the costs of energy are as follows (at 1980 prices): two energy sources, 328 MSkr; three sources, 340 MSkr; four sources, 345 MSkr; and five sources, 350 MSkr.

Renewable energy sources for powering the production of activities such as agriculture, forestry, and fisheries have been important for centuries to the make-up and viability of Gotland's economy. Recently, interest has also focused on returning to the use of renewable energy sources for the direct production of fuels and electricity. Viable renewable energy technologies in the form of forest plantations, straw, wood wastes, solar heating, and wind-electrical and wave-electrical systems were considered. Earlier rough calculations with conservative assumptions (see Table 2.10) indicated a potentially important role for their contribution to the energy supply. More precise calculations, as well as modeling efforts (models 2 and 5 in Chapter 3), also indicated that renewable energy technologies could provide substantial amounts of required energy. Under the objective of maximizing local energy production (see Figure 3.14), all electricity could be supplied by wind and wave power, and well over 50% of the fuels via biomass. If required energy is reduced through increases in efficiency, then renewable energy technologies could potentially provide all energy needs (see Figure 3.15). This is also possible if the amounts of water and fertilizers available for renewable energy technologies are increased (see Figures 3.16 and 3.17). However, these results lead to a more expensive energy system than could be obtained by importing fossil fuels and electricity over the mainland cable. The objective of minimizing total energy cost leads to a dominance by imported fuels and electricity (see Figure 3.14), but rising prices of coal and oil would lead to declining imports and a greater importance of renewable energy sources (see Figure 3.18).

It should be emphasized that these results are all theoretic and based on current best estimates of costs, production and resource requirements. Assumptions have been conservative so as not to overestimate the potential contribution of alternative technologies. More confidence in such analyses can be generated by using data from currently operating prototype systems. The wind-electrical prototypes operating on Gotland will certainly provide more accurate information on these systems in the near future. It would also certainly be important to obtain detailed field results for energy plantations, especially with regard to irrigation, fertilization, and water quality impact, since this technology is such a large potential contributor. However, if it is anticipated that future costs of coal, oil, and electricity from the mainland will rise, then it seems germane to seriously consider the potential contribution of renewable resources.

In a more general sense, we can look at the impact of primary inputs on the economy by using the input-controlled, input-output formulation discussed in the "Input-Output Models" section in Chapter 3. The row sums of the matrix

$$(\underline{\underline{I}} - \underline{\underline{\bar{A}}})^{-1}$$

Table 4.5. Total Impact on Output Throughout All Sectors for a 1-Skr Change in Primary Inputs for the Indicated Sector

Sector	Impact on Total Output (Skr)
Animal	2.48
Crops	2.28
Forestry	1.31
Fishery	1.04
Mining	1.03
Slaughtery	1.08
Dairy	1
Sugar	1
Other food	1.5
Wood	1.23
Printing	1.07
Stone and soil	1.04
Communication equipment	1
Other workshops	1.37
Electricity and water	1.4
Construction	1.53
Wholesale	2
Transports	1.91
Tourism	1
Other services	1.79

will give the total effect on output, throughout all sectors associated with a 1-Skr change in primary inputs for sector i. We calculated the above matrix by using the intersectoral flows in Table 3.2 to arrive at the row sums. Table 4.5 indicates, for example, that a change of 1 Skr in the animal sector would produce a 2.48-Skr change in total output. It is seen that sectors such as animal, crops, wholesale, transports, and other services have rather large impacts (these numbers refer to increases or decreases, depending on the change in the primary input).

Hydrologic Considerations

Water availability and groundwater nitrate pollution have figured importantly in our study. The extent to which water and environmental quality may become limiting factors for particular economic activities will be decisive in Gotland's economic future. Analysis of water requirements per Swedish crown of economic output from different sectors (Table 3.5 for model 1) indicated the high values for animal production, slaughterhouses, dairies, tourism, and stone and soil. Other sectors were relatively low in water demand, such as crop production, workshops, or fisheries. Potential water pollution problems might arise from high loads of organic wastes associated with animal production, slaughterhouses, or dairies. Furthermore, nitrogen oxide generation is high in such activities as crop production, fisheries, transportation, and tourism, at least on a per unit output basis.

Models of water supply and quality revealed some interesting results. With water as a constraint in a linear programming problem that maximized value added (model 3), the importance of water to total gross regional product (GRP) was evaluated (see Figure 3.6). There are decreasing returns to scale as water becomes less limiting; increasing water availability by modest amounts above existing levels could allow substantial increases in GRP assuming a demand for the increased output. The calculated shadow price for water is substantially above the existing price, which suggests that water is undervalued. The effect of raising water prices on the price of final goods is small (see Table 3.13). Raising water prices might lead to more efficient use without substantially impacting on the economic situation.

Available water was a significant parameter in the optimization models for the energy system (model 5), and increasing water availability was crucial to the development of energy plantations for energy production (see Figure 3.16). The simulation model of water and nitrate (model 6) established the relationship between fertilization rates and nitrate levels in groundwater. This relationship turned out to be essentially linear (see Figure 3.22), and it served as information for the energy system optimization model (model 5). Modification of fertilization regimes from existing conditions could enhance plant growth rates and decrease groundwater nitrate concentrations. For sugar beets on glacial till, for example, distribution of fertilization during the growing season resulted in lower groundwater nitrate concentrations and greater uptake of nitrate by the plants.

Nitrogen runoff from land and into the Baltic Sea was found to be both significant and on the order of nitrogen fixation by blue-green algae in the coastal zone (see Table 3.21). The models of the coastal zone, which did not include turbulent mixing and dispersion of nitrogen, indicated that the biological processes were able to assimilate the load of nitrogen runoff. In fact, this nitrogen was an additional energy source to the coastal system, enhancing production as well as fish stocks.

Water and the Landscape

Water is obviously important to the ecology of the landscape. Since the extensive drainage of the nineteenth century, the landscape has been extensively modified. Highly productive and diverse wetlands have been replaced by simpler, managed agricultural systems. These agricultural systems, in turn, have been productive because of the fertile organic storages that had accumulated in the soils. Drainage and the breakdown of these peat soils have contributed to high nitrate concentrations found in both ground and surface waters. The island, as a whole, would have a very different character if drainage had not occurred, if water storages were higher, and if wetlands occupied a greater percentage of the land, although at the expense of agricultural land. The dry period during 1983 emphasized the extent to which drainage has altered the hydrologic regime of the island, with creeks drying up and water levels dropping. It seems that the water supply situation on Gotland has become too dependent on yearly precipitation and that more long-term climatic trends should be considered. If a period of more frequent dry spells emerges, then restrictions on water use would probably be necessary. This would lead to a decline in economic output and would generate incentives for water management, such as the reestablishment of wetlands for water storage, stream maintenance, and wildlife habitat.

Costs and Benefits of Preserving and Restoring Wetlands

The drainage of wetlands in the nineteenth century made sense from the human point of view. Agricultural productivity was enhanced, and any groundwater quality problems generated were neither noticed nor considered serious because of low water demand from the small island population. In the language of economics, the benefits of using the rich organic soil storages far outweighed the costs of society neglecting the impacts associated with the destruction of the wetland ecosystems. However, the present situation is different. The peat soils in the drained areas have disappeared to a great extent. Water quality has deteriorated in areas where there is now a significant demand for water. In addition, the growth of society and the economy has increased the demand for water of sufficient quantity and quality by activities other than agriculture. From a regional perspective, it might make more sense now to reestablish wetlands in some areas of the island. This would reduce the amount of arable land engendering a cost. On the other hand, increased water supplies of adequate quality would allow irrigation in agricultural areas and the support of other economic sectors. Whether the benefits would outweigh the costs is a subject that requires evaluation of real situations on the island. We suggest that a periodic flooding and drainage of different areas might even be a long-term strategy for maximizing output. The wetland areas would build up organic soils and maintain water supplies. After sufficient organic soil accumulation, they would be drained and used for agricultural production, whereas other areas would be reestablished as wetlands to maintain sufficient water storage on the island. Alternative strategies could be evaluated regarding the relative cost and benefit implications. Assuming that the benefits to the region did outweigh the loss in agricultural output, the farmers could be paid to reestablish wetlands in designated areas.

Economic Development and Resource Considerations

Depending on which economic sectors grow, changes will take place in the economy, resource use, and environmental impact. Information concerning these areas of interest can lead to assessment of anticipated investment in various sectors of the economy, with an eye to encouraging those activities that will be considered most beneficial or most likely to succeed in the future. There is obviously a notion of trade-offs involved here. For example, some activities may be low in generating value added or employment, but high in terms of energy consumption or pollutant generation. High energy-consuming activities might be considered undesirable, because they might not be sustainable in the future. Activities that generate high levels of pollutants would also be considered undesirable if the environment was impacted on to an unacceptable degree, perhaps leading to an eventual deterioration of the carrying capacity of the region. Such potential impacts might also stimulate the development of less polluting technologies or investment in pollution controls.

Theoretically, trade-offs among economic impact, resource use, and environmental degradation could be qualitatively imagined as a three-dimensional space (Figure 4.2). If a number of economic, resource, and environmental measures were considered, then an n-dimensional space could be defined. High levels of resource use or environmental degradation could only be acceptable if the economic benefits were high. This defines an increasing function that could be represented as a surface; points lying above this surface are acceptable, while economic activities lying below this surface would be less desirable. Of course, this is a very ideal picture. In reality, a close look at the expansion of given economic activities would have to be undertaken to make a judgment about desirability and impacts. For example, referring back to Table 3.5, the communication equipment (workshops) sector has low energy, water, and environmental impacts, and high value added per unit of output. In contrast, dairy has relatively higher energy, water, and environmental impacts, with more modest value added. If the choice were available, expansion in the communication equipment sector would be preferable according to its position in Figure

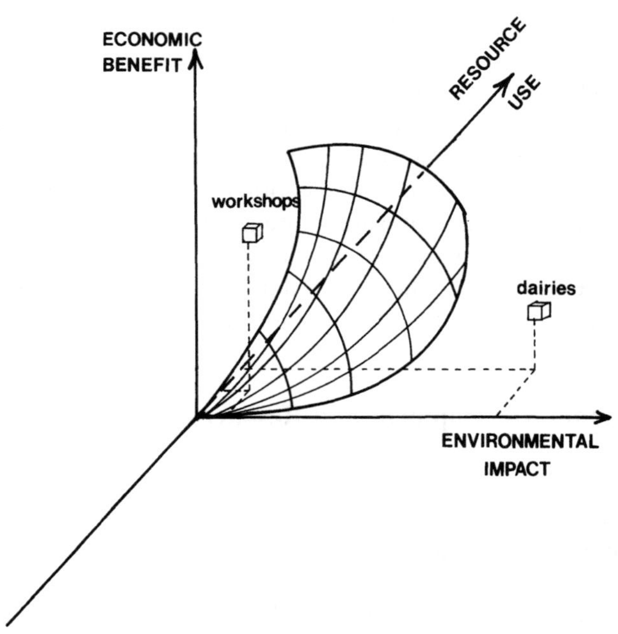

Figure 4.2. Diagramatic representation of trade-offs among economic benefits, environmental degradation, and resource use. Different sectors will lie at different points in this space. A surface separating acceptable from unacceptable activities can be imagined that represents the notion that activities with high ecologic or resource impact should produce high levels of economic benefit.

4.2, which is somewhere above the imaginary surface; whereas sectors such as dairies or stone and soil lie below. Thus, a sector such as stone and soil would generate more environmental degradation and resource use for the economic benefit accrued in comparison to communication equipment.

Model Results and Resource Management

Many of the modeling results and their implications for Gotland have been discussed in the preceding paragraphs. Although it is true that the models formulated in this study can only answer a few questions of concern to planners, they nevertheless address some of the most important issues for future development. In addition, they constitute a reasonable foundation for continued modeling efforts of problems of more immediate utility. It should be remembered that our models could be made more precise if the funds were available to collect data needed for parameterizing such models. We had to work within the confines of a limited set of data and with little funding for detailed investigations. Therefore, we tried to maintain our perspective, as much as possible, at the level of the region as a whole.

We certainly think we have reached beyond knowledge existing prior to our study to an understanding of interrelationships on Gotland. The resource use calculations using input-output analysis (model 1), as well as our analyses of resource use in the various sectors, provide new information on the total impact of expanding activities in existing sectors. For example, Table 3.5 indicates for a unit expansion in economic output that agriculture, fisheries, and wholesale trade will generate the most total employment. If we want to invest in sectors with high employment multipliers, Table 3.6 points to areas such as slaughteries, dairies, and tourism. The economic optimization model (model 3) in Chapter 3 focused on the impact that water had on the total economic output of the island, as well as on the effect of price changes in important goods. The results indicate that it takes very large changes in prices of water to substantially affect prices of final goods (see Table 3.13). However, prices of final goods are rather sensitive to changes in oil prices. The shadow prices obtained for water from the optimization model were much higher than existing water prices. These results for water indicate that the planning authorities might try to experiment with higher water prices to induce conservation and more efficient use.

Several models of the energy supply system were formulated that show interesting relationships to the economy (models 2, 4, and 5 in Chapter 3). From Model 2, we see that even modest increases in oil prices will make it practical to introduce energy plantations on the island at a fairly early date. Since the optimization model maximizes value added, sectors such as food industries, agriculture, and workshops are relatively insensitive to the rising oil prices and also are responsible for the bulk of the output. Thus, during periods of energy interruptions, these sectors would be best for the allocation of energy. Model 4 in Chapter 3 shows the importance of not only diversifying the supply of energy, but also distributing it across sectors. Again, allocation to the most energy-efficient sectors can substantially minimize potential disruptions in total economic output. The detailed analysis of renewable energy production (model 5) indicated the extent to which indigenous energy sources could meet demand, although at an increased cost.

The simulation models dealing with hydrology and water quality (model 6) were as detailed as existing data allowed. These simulation models were used to estimate the impact of different regimes of irrigation and fertilization. Thus, a planner could use such models to estimate the effects of agricultural expansion or the establishment of energy plantations on water quality, water quantity, and crop production. We used such a model to estimate the effect of land-use change in the area of Lummelunda on water quality. The coastal ecosystem model was an example of a detailed simulation model that could be used to estimate existing or potential impacts. The model results indicate that present loadings of nitrogen runoff from land can be assimilated by the biological community. The model could be used to do an impact analysis. For example, in an earlier version of the model a simulation was done on the impacts on the coastal ecosystem caused by an expansion of fish-processing capacity on the island and increased herring and cod fisheries. The expansion also generated a set of interactive impacts within the economic system. The increased capacity of the fish-processing industry allowed larger catches. A tripling in cod and a 150% increase in herring catches were simulated by the model, which indicated a tolerable response of the coastal ecosystem. This increased harvest resulted in an income gain for herring of 136,000 Skr yr^{-1}; and for cod, 321,000 Skr yr^{-1}. On the other hand, the input-output multiplier coefficients (see Table 3.5) showed that this expansion would result in an increase in resource use, pollutant outputs, employment, and value added, as summarized in Table 4.6. Thus, a combination of ecosystem simulation models and input-output models allowed a system impact analysis of an increase in fisheries. Similar types of combined

Table 4.6. Expected Impacts for Increasing Cod Harvest by 3 Times and Herring Harvest by 1.5 Times[a]

	Change[b]	Percent of Total[c] Economy
Employment	11.4 persons	0.044
Oil consumption	3.5 TJ	0.032
Electrical consumption	32 GJ	0.0028
Water	100,000 dm^3	0.15
BOD	46 kg	0.00076
SO_2	503 kg	0.023
NO_2	45,531 kg	0.25
Value added	237,640 Skr	0.015

[a] Change in value of output for increased herring and cod harvest approximated to be 457,000 Skr for 1975.

[b] Change = 457,000 Skr multiplied by corresponding multipliers in Table 3.37.

[c] Compared to total levels of economy in 1975; 1972 figures used for BOD and water (see Table 3.37).

economic and ecologic assessment could be conducted in a similar fashion for other changes in the region.

Some of our models can be used to suggest trends or potential optimum solutions for economic development, while others were designed to answer more specific questions concerning environmental impact. However, it is obvious that in a real planning situation, almost all of the models have to be modified or reformulated to cope with emerging issues or problems. In any case, an updating of the data set will be needed if the models are to be used as a basis for decisions. General discussions of models, management, and regional analysis are contained in Chapter 5.

Site Specific Applications: The Area of Lummelunda

As previously alluded to, part of the research effort during the latter years of the project focused on a site specific study of the Lummelunda creek run off area about 20 km north of Visby. A detailed study of a particular area was thought to be more relevant to the needs of planning authorities. The results of these investigations are not the subject of this volume, but several points are worth mentioning (Nilsson 1982, Nordberg 1983). Remote sensing techniques were used to chart the course of land-use change, with the ultimate objective being development of a technology for an efficient environmental overview of the entire island. However, the time and manpower required for only the small study site was so excessive that we realized the futility of rapid analysis of the entire region. Nordberg (1983) was able to evaluate the extent of land-use changes, such as clear-cutting and replantation, and the reduction of natural wetlands and deciduous forests from 1958–1980. The other major thrust of this analysis was to determine the relationship between land use and water quality. The hydrologic model (model 6) presented in Chapter 3 was used by Nilsson (1982), along with field measurements, to assess the state of water quality and quantity in this area. This was an important demonstration that our models could be used in a manner in which specific information relevant to the planning authorities could be generated.

Investment Ratio and Regional Carrying Capacity

There have been several hypotheses relating the competitiveness of a region to renewable energy flows and sustainable carrying capacity (Zucchetto 1975b, Odum and Odum 1976, Odum 1983, Jansson 1985). Referring to the generalized diagram of a region (see Figure 1.3) there is a flow of embodied energy in imports and exports, as well as work accomplished by the renewable energies in the region. The investment ratio is defined as the ratio of imported energy to natural energy flows, all energies are expressed in equivalent units of quality. The investment ratio can be thought of as a measure of development; the higher it is, the more fossil fuel-dependent the economy. This ratio can also be interpreted as a macromeasure of carrying capacity. Regions with high ratios have significant human impact, and the lower energy levels of the natural systems reduce their ability to compete. In the long-run, the sustaining capacity of the natural systems may be eliminated. On the other hand, regions of low investment ratio would be relatively undeveloped and presumably able to attract investment and sustain additional impact. Furthermore, regions with low investment ratios have more services provided by the natural environment. All other things being equal, Gotlanders should be able to sell their exports at a lower price than regions with higher investment ratios (Zucchetto 1975b).

Using the approximate energy quality factors developed by Odum (1983), the flows of energy are converted to equivalent units in Table 4.7 to make a rough estimate of the investment ratio. The investment ratio turns out to be about 0.2 (141,606 ÷ 697,400), which is a relatively small number compared to 2.5 for the United States (Odum 1983), 1.7 for Sweden (Limburg 1980) or 13 for Miami-Dade County in Southern Florida (Zucchetto 1975a). Based on this notion, therefore, Gotland would seem to be in a competitive position and also be able to absorb or attract more development. Although this measure of carrying capacity seems to make some sense, it is perhaps too macroscopic a measure to capture realities of detailed occurrences. For example, in the nineteenth century, the natural productivity was not much different, but human activity was far less. However, the extensive cutting of forests led to a crisis situation (Jansson 1985). This may be a good example of a threshold effect, in which existing conditions may be tolerable, but could lead to a decline of the total system. Too much forest cover being removed could lead to extensive water and wind erosion and to an eventual overall decline of the total system. This gross measure of carrying capacity does not seem to pay enough attention

Table 4.7. Purchased and Natural Energy Flows for Gotland, 1972

	Heat Equivalents (TJ)	Solar Equivalents (PJ)[a]
Natural ("free") energies		
Gross photosynthesis	100,000	92,000
Wind energy	152,000	47,880
Wave energy	56,000	514,080
Free energy of salt/fresh water	6,000	41,400
Potential head of water	200	2,040
Totals	314,200	697,400
Purchased energies		
Imported fuels	13,817	93,956
Imported electricity	540	14,610
Embodied energy in imported goods		
Fuels	2,227	15,144
Electricity	655	17,816
Totals	17,239	141,606

[a] Global solar calories/calorie are given as: wind, 315; gross photosynthesis, 920; fossil fuel, 6800; chemical potential of water, 6900; electricity, 27,200; waves, 9180 (1.35 × 6800); potential head of water, 10,200 (1.5 × 6800). Conversion factors are obtained from Odum (1983), p. 252.

to resources in limited supply. More development on Gotland will be constrained by limitations of water. Furthermore, its distance from mainland centers will probably discourage investment and growth that might have occurred if its location were more fortuitous.

Future Design for Gotland: What Does History Tell Us?

From the historic description in the preceding sections, it is clear that Gotland has changed from a subsistence economy based on local natural resources to an open economy affected by events on the Swedish and foreign markets. The society on Gotland has become strongly dependent on a steady inflow of energy, goods, and money from outside to maintain its structure and production. Urbanization has led to a concentration of people in Visby. The ratio of rural-to-urban populations dropped from almost 10:1 during the middle of the nineteenth century to 0.9:1 at present (see Table 2.1). This ratio might be comparable to a production-to-consumption ratio of an ecosystem that is high in a young system, close to 1 in a mature steady-state system, and below 1 in a net consuming system running on imported material. The rural-to-urban ratio seems to have been below 1 once before in Gotland's history; namely, during the Hanseatic period of the Middle Ages. At that time, the population in Visby received most of its income from foreign trade. Later came a long period of decline, when the island was cut off from the outside support and forced to rely on its own resources (Jansson 1985). In the same way, the present society is extremely vulnerable to events disturbing the transports and exchanges between Gotland and the outside world. The politicians have made no serious attempts at easing this dependence by making the region more economically self-sufficient. Nor has there been much effort to make secure the productive potential of the natural resources. The transformation of natural ecosystems and the exploitation of land and water resources have been allowed to take place without any evaluation of their combined effects on the functioning of the regional system. The transformation of land has been going on since humans came to Gotland. Most of the island's present appearance can be regarded as a cultural product. The most considerable changes, however, have occurred during the last 2 centuries. The major transformation has been due to the almost fivefold increase of arable land, which has significantly reduced the area of deciduous woods, meadows, mires, and lakes. Conifers seem to have reforested the area lost during the extensive exploitations during the eighteenth and nineteenth centuries, while some forested heathlands have probably disappeared permanently.

What has this transformation of land meant to the overall functioning of Gotland's collective ecosystem? One way of estimating this impact would be to compare the total amount of solar energy fixed in gross photosynthesis before and after the transformation took place. From reasonable assumptions about past productivities it seems that the present gross photosynthesis of the whole terrestrial system is lower than that of 1700 A.D. Furthermore, the present production of agriculture depends on an input of extra energy in the form of fossil fuels, fertilizers, pesticides, and machinery. This amounts to about 1.3 PJ/yr of fossil energy. If this extra input were to disappear, then the productivity of the agricultural system would most probably drop to a much lower value. One reason for this is the decreased content of organic matter in the soils. The historic production was low due to a technologically primitive cultivation technique. However, the maintenance of the organic contents of the soil was provided for by the application of manure from a high number of cattle (see Table 2.1). The number of cattle per km^2 of arable land decreased from 133 in 1800 to 38 in 1900. Then the use of imported fertilizers began, making manure less important in relation to the demand for dairy products and meat. The field area is so large that the application of manure only recycles about 20% of the energy in organic matter removed by the harvests. However, the most substantial loss of organic soils has been through the oxidation of peat on drained mires. In certain cultivated mires, the peat has been found to decrease about 1 cm/year after the groundwater level has been lowered (Nilsson 1982). The average depth of peat was 1–2 m around 1900 and is now about 0.3 m (Jansson and Zucchetto 1978a). If this decrease is assumed to be valid for the entire area of drained wetlands, then the island has lost about 7700 PJ of energy stored in peat during the last hundred years. This is equivalent to several 100 years of present fuel consumption.

Another environmental impact caused by agriculture is the increasing amount of nitrates in groundwater. The average content of nitrate in groundwater is now 8 mg/liter (Spiller et al. 1981). Drainage was instituted to increase runoff of water from agricultural fields. However, this has decreased the water storage capacity of the soil to such a degree that a shortage of water is experienced in the summer months. At present, water seems to be the most decisive limiting factor for production both in natural and managed terrestrial ecosystems on Gotland (Jansson and Zucchetto 1978a). Investments are now being made to build freshwater reservoirs and irrigation systems to replace part of nature's own storage capacity and ability to buffer the lack of rain during the growing season.

The competition for water between different sectors in society has just begun. This will certainly be one of the major issues for the future. Two activities that are not thought to have any considerable impact on freshwater reserves are limestone quarrying and cement production. These large-scale activities on Gotland have a negative impact on the potential groundwater build-up by increasing runoff. In fact, the consumption of fresh water in cement production has been found to be roughly 2 liters/Skr of economic output if both direct and indirect consumption are taken into account (Andréasson 1984). Next to food production, this is highest among the economic sectors on Gotland. Regarding limestone itself, the total annual exploitation amounts to approximately 200 million tons. When compared to the natural losses due to weathering and erosion, the human exploitation is 200–250 times larger. However, Gotland's upper layers of limestone are continuously supplemented from further below the ground by means of land uplift, which represents an even larger flow.

Many of the developments on Gotland during the last 2 centuries have been shown to imply a deterioration of the capital of natural resources that had been built up during the 10,000 years since the island came into existence. On the other hand, the modern structures of society represent a new set of values that may, for at least a period, keep up the productivity by attracting necessary energy and capital from the outside. Also, recent history seems to tell us that new techniques allowing a more effective use of available resources have always appeared; it would consequently be unwarranted, from an economic point of view, to be saving for the future by placing restrictions on the use of natural resources. However, it can also be seen from history that the introduction of new technologies has increased the total energy consumption in the society. This is true even for the recent improvements in technology in both cement production and the food industry on Gotland. These have significantly decreased the energy consumption per unit of output; but at the same time, they have forced a considerable expansion of the production, which in turn has led to a total increase of the energy consumption on Gotland. Instead, the gains from the more effective uses of energy have been obtained in the national energy budget. This is one example of the fact that Gotland is no longer an isolated system, but is an integral part of the Swedish economic system.

Recommendations to the Planning Authorities

During the course of this research project, efforts were made to transform some of our results into useful information for the planners of Gotland. Since the project, especially in the early phases, was primarily a research endeavor, significant time or resources were not spent in developing planning methodologies. There were several presentations and interactions with the planning authorities on Gotland to communicate our approaches and results. Based on the accumulated experience and analysis resulting from the Gotland project, as well as from other work in the area of environmental systems approaches, the following general recommendations are suggested for the planners on Gotland. These suggestions do not require direct adoption by the planning authorities themselves. At the very least, it is hoped that incentives would be established to generate inquiry in the areas outlined.

(1) Diversify the energy supply system. The strategy of "not putting all your eggs in one basket" is a familiar one. Certainly, the simulations of the effect of energy supply diversity on economic output reemphasizes this strategy. Further studies should be implemented to calculate the economic and environmental costs of diversifying the energy supply system against the benefits of the economic stability it generates.

(2) The potential use of land for "energy plantations" should be investigated further. What land area is realistically available, what productivity levels can be expected, what are the trade-offs if agricultural land is used, and what are the environmental impacts? These are the main questions requiring more specific answers.

(3) Means of increasing energy efficiency and using low-quality heat should be investigated. Obviously, there has been much emphasis during the past 10 years on increasing the efficiency of energy use. Governmental subsidies and

tax incentives were available. In addition, higher prices for energy acted as an economic incentive to conserve. However, the potential for energy savings on Gotland for existing technologies and fuels may have been realized already during the past few years. We suggest that there may be new innovative ways for using what is now "waste" heat from several activities. Obviously, using waste heat from factories or power plants for local heating in buildings is possible. It may also be efficacious for enhancing the productivity of managed ecosystems with this energy source; e.g., using waste heat from factories to provide a productive environment for aquaculture may be one area worth investigating.

(4) Maximize the services rendered by the ecologic and environmental systems. Strategies for using the products and services of the biosphere in sustainable and efficient ways imply considerations of environmental planning, ecologic engineering, and resource management. The coupling of waste heat with fish ponds cited in recommendation 3 (above) is one example. Additional examples that already exist or should be investigated are: (1) the use of fish-processing wastes as an energy (food) source to feed minks for the production of high-quality furs, (2) release of salmonid fish and eel into streams and near-shore areas for their eventual harvest after returning from migratory journeys at sea, and (3) the restoration of some of the drained areas to wetlands for purposes of water conservation. This last suggestion involves a trade-off between loss of income from existing land use against maintaining an adequate supply of water for human and industrial activity on the island.

(5) Water quantity and quality investigations should be continued. Our efforts, as well as those of others, have dealt with the problems of water on Gotland. Better data are needed on groundwater supplies, consumption by various activities, and human impacts on quality. As mentioned in recommendation 4 (above), the management of wetlands for water conservation may be part of an overall water management strategy. As the economic optimization models indicate, the total economic output is maximized under higher shadow prices for water. It is suggested that higher water prices should be investigated as an incentive for increasing water conservation, which requires no economic sacrifice for moderate increases in water prices. An industry-by-industry analysis probably should be undertaken by assessing the economic and technical adjustments under conditions of higher water prices.

(6) Alternative fertilization regimes should be investigated for agricultural crops. Our simulations indicate that the timing of fertilization influences both nutrient plant uptake and losses due to runoff and infiltration to groundwater. Experimental work, as well as mathematic modeling, can lead to further understanding of the establishment of optimum fertilization strategies that maximize plant uptake and reduce water quality impacts.

(7) Develop plans for economic diversification and employment-generating economic sectors. This has already been a policy of the central government in terms of its establishment of incentives for industrial location, especially in the early 1960s. Our study has shown that the relationship between energy and job creation is very much a function of which economic activities are in operation.

It is suggested that renewable-based technologies and resources should be investigated for their potential for employment generation, especially on a seasonal basis.

(8) The efficacious integration of systems analysis and modeling into the planning process will probably take place only if a dedicated group of systems modelers is established to work in close connection and on a day-to-day basis with a planning agency. We suggest that such a group be established to serve the planning authorities agency, either directly on Gotland or via communication line to an appropriate department at a university. This group could contribute to the solution of problems that are of immediate concern to the planners. At the very least, frequent workshops should be held with systems analysts and planners to demonstrate what systems analysis can or cannot contribute to the planning process.

(9) Establish a comprehensive environmental and resource monitoring system on the island. In our work, there were many instances in which sufficient data were not available to generate conclusive answers to questions of interest. Firm-by-firm data for resource use as well as air, land, and water impacts would give an important detailed view of environmental impacts. Other relevant data requiring more refined collection include spatial estimations of land-use change and related water quality and quantity aspects, estimations of soil erosion, and more data for the coastal areas regarding the impact of expanded fisheries and resource extraction.

(10) The energy optimization models indicated that it was feasible to meet a large proportion of energy demand with indigenous renewable resources. However, this strategy would be more expensive than using fossil fuels and imported electricity. This issue should continue to be addressed as technologies emerge. Additionally, electricity from the mainland may not be economically competitive within 20–25 years if nuclear power plants are discontinued and proven to be a nonviable technology.

(11) A cost-benefit analysis should be conducted for the reestablishment of wetlands in agricultural areas. The cost of lost agricultural output should be weighed against the advantages that would accrue for increased water supply and quality.

Plate 4. Rapid runoff from the Lummelunda creek during spring (Photo, T. Hilding).

5. General Considerations for Regional Studies

This last chapter is an attempt to distill the experience of the Gotland study into a set of general observations that apply to all regional ecologic-economic studies. There were many accomplishments associated with the Gotland project. However, there were also a number of frustrations and shortcomings that illustrate the general difficulties encountered in conducting geographically large-scale studies that attempt to synthesize and incorporate both the natural and manmade environments of a region.

Comparison to Other Studies

There are, perhaps, hundreds and possibly thousands of regional studies that have been undertaken with various projected aims. We have picked a few that are illustrative of different schools of thought or are particularly relevant to island systems. Some of these studies have been important to UNESCO's Man and Biosphere (MAB) program in an attempt to deal with integrated problems of humans and nature. We try to indicate the differences that exist between the various projects and our own and to point out those areas where strengths or weaknesses of our project emerge.

The South Florida Study

Since much of our motivation for the Gotland study originated with the south Florida study (Browder et al. 1976), there are a number of important similarities between the two. The major theme in the Florida study was to seek a balance between humans and nature; from the very beginning of this project, attention was paid to the interactions of ecosystems and economic systems. The emphasis was on understanding the contributions of the natural systems to the region of south Florida and the importance of development occurring in a manner that would take advantage of the functioning of the natural systems.

Energy was a most important part of the study because of its role in powering both the human economic system as well as the natural processes. Detailed arguments were made for the case that the United States was entering a period of less energy availability. This would promote strategies by society to make better use of the region's capacity to provide vital services through natural ecosystems. Macroregional analyses were conducted for the economy, the use of energy, and the natural resources. Detailed land-use maps were constructed to delineate the changes that had taken place from 1900–1973. Energy-economic and hydrologic simulation models formed an important part of the analytic efforts. There was much attention focused on water because of problems in South Florida. Land-use changes, drainage, salt water intrusion, and expanding developments have all led to a severe water crisis in south Florida. Simulation models, which focused on energy and the economy, were formulated to assess the impact of various development strategies. The use of wetlands for providing services to humans, in the form of tertiary sewage treatment or water recharge, was an important aspect of the study.

In light of this short review, the Gotland study was similar in many ways to the south Florida study; however, there were important differences. Although there was much interest in the impact of declining energy supplies on the economy in the beginning of the Gotland project, this was deemphasized to some degree as the project moved along. The notion that there might be a general decline was too vague as to the extent or precise time of its possible occurrence. Nevertheless, possible shortages of energy still dominated many of our studies. There was also the fact that the energy disruptions of the 1970s produced extensive economic adjustments, many of which had been realized by the early 1980s. The emphasis on energy dwindled, even though the continuing economic problems could be traced, in part, to the energy disruptions of the 1970s. The south Florida study ended in 1976, when the emphasis on energy was more topical. Rather than dealing only with simulation models for the total economy, we shifted our focus towards optimization models that could at least deal with the problem of how to "best" allocate energy among different activities in the economy. This was a focus that was not central to the south Florida study.

The Gotland study also entered into more detail regarding the economic system. In addition to macroeconomic analysis, as in the south Florida study concerning energy and economy, we also devoted much attention to sector analysis.

Details by industrial sector concerning energy, economics, labor, and pollutant generation, as well as the economic flows among economic sectors, were not part of the south Florida study. Our emphasis on using input-output analysis for the economic system was also quite different. Both studies used simulation models for the ecologic and environmental systems. Finally, both studies tried to emphasize the work done by natural ecosystems by virtue of the fixation of solar energy. However, we also spent a good deal of effort attempting to assess the potential contributions of alternative energy technologies to Gotland's economy. Our emphasis on optimization methods, regarding the energy supply system, thus differed markedly from the Florida study.

Hong Kong Human Ecology Program

The design of the Hong Kong study, which addressed problems of a highly urbanized area, differed in many ways from that of the Gotland study. However, both studies stressed the importance of energy in "the modern industrial phase of human development" (Boyden 1979) and the need for an integrated ecologic approach to understanding global and local problems. Both studies emphasized that the properties of a system as a whole must be looked at. In the Hong Kong study. much attention was focused on the life conditions and well-being of individuals as they were influenced by the total environmental impact on their own personal environments. Studies of individual behavior patterns and perceptions of the "biopsychic state" were important parts of the project. In our study, the social conditions were not subject to any special investigations based on the judgment that extensive Swedish income equalization and social welfare programs have helped to equalize any large differences among social groups or Swedish regions. Instead, we put more emphasis on comparing economic sectors as to their impacts on such important social factors as employment, income, and health.

A rather detailed energy analysis of the so-called extrasomatic energy (not expended through metabolic processes within living organisms) was made for Hong Kong. The analysis included outputs of energy supplies, as well as the use of energy within the city and its side effects; air pollution models were used to predict atmospheric pollutant concentrations. Differences in household energy use among socioeconomic groups, as well as impacts of energy on life conditions (such as greater centralization in industry as a result of greater energy use), were also considered. The energy study of Gotland went beyond that of Hong Kong in several ways. In the first place, we included a comprehensive analysis of the natural energy flows and how these interact with the "extrasomatic" energies to produce economic output. We also paid more attention to the relationship between economic activity and energy use; the Gotland study emphasized this aspect with both energy and economic statistics.

The use of modeling techniques to detail the various environmental and ecologic consequences was more important to our study. For example, a water balance was conducted both for Hong Kong and Gotland, but we additionally developed ecologic simulation models to assess the impact of human activity on water quality and quantity; and, we incorporated water considerations and constraints into our economic models by means of input-output and optimization techniques. We think that by means of such integrated analyses as resource use, economic activity, and environmental impacts, we were able to encompass the total system of humans and nature in a more comprehensive way than did the investigations of the Hong Kong study.

The Case of Eastern Fiji

A study of the population and environmental relationships for the tropical islands of eastern Fiji is of special interest because it focuses on problems of island ecosystems (Brookfield 1980). These islands, being small and isolated, are quite vulnerable to disturbances. The study was concerned with the management of environmental resources, the impact of external forces, and the introduction of exotic species. For example, examined were the effects of short-term variability associated with drought and hurricane, the impact of mining on erosion, and processes at the land-sea interface affecting coastal geomorphology. The investigations took mainly a physical geography approach. Demographic analyses were conducted and population processes were seen to be different from those of large populations. Much emphasis was put on notions of carrying capacity. Although this concept is somewhat ambiguous for human populations (see the section on "The Issue of Carrying Capacity"), it is easier to deal with in economies that are close to subsistence levels. For example, one can evaluate both the potential or existing production of terrestrial and marine resources and a corresponding estimate of the size of the human population, which can be supported under different management schemes. Resource analyses were conducted for agriculture and estimates were made for the living marine resources of the south Pacific, including present and potential exploitation. For one of the islands, Taveuni, linear programming techniques were applied to the local economy.

The island of Gotland is much larger than those of eastern Fiji, and it probably is not as vulnerable to outside forces. Furthermore, Gotland is not so far removed from urban areas on mainland economies as are the Pacific islands, which facilitates important exchanges of energy, people, goods, and information with larger economies. Nevertheless, our study had several things in common with that of Fiji. Both studies focused on resource use. The energy analysis of the less-developed subsistence economy of Fiji only considered the production of food and the use of human labor, whereas our study also encompassed the detailed flow of imported fuels and electricity. Natural energy flows, apart from photosynthesis, were not considered in any systematic way for their potential support of the Fiji islands through alternative energy technologies. The more diversified economy of Gotland required more extensive economic analysis of the various sectors. Except for the one island of Taveuni, mathematic modeling was not significantly used in the Fiji project. In fact, one serious weakness of the study seems to be that different analyses were conducted for different islands. The advantage of the single-island study of Gotland was that the use of natural resources and the environment could be linked to the economy for the same island. We think that the Fiji island study could become more important by considering all the islands together as one system, with attention to the interactions that take place among them and with the surrounding sea.

Ecologic-Economic Analysis for Regional Development

The discipline of regional science has been concerned with the study of regions (Isard 1975). The main emphasis of this field has been an understanding of the spatial organization of economic systems, with special emphasis on economics and mathematic methods. However, environmental considerations have been introduced into the regional science literature and are exemplified by Isard (1972) in his book: *Ecologic-Economic Analysis for Regional Development*. As Isard indicates in his preface, this book differed greatly from his previous works in showing the intricate interrelationships between the economy and the ecosystem. In spirit, therefore, Isard's approach to regional studies was certainly similar to ours in its basic theme. His emphasis, however, was not on energy per se, although there is no reason, in principle, why energy could not have been included. The applications cited in the book, for example, are related to impacts of marinas and resource recovery; i.e., on rather limited sites, but its theoretic arguments could be applied to any-sized region. Essentially, the methodology is to consider a matrix of economic and ecologic interactions; each, in turn, is divided into land and marine interactions. Thus, there are interactions between activities in different sectors of the economy and in the ecosystems that generate impacts on commodity flows. These are listed in a matrix as to the level of commodity output or input per unit output of the activity. The principal idea is the linearization assumption, in which inputs and outputs to a given activity are assumed to be linearly related. This is recognized as a limitation of the method. On the other hand, making linear assumptions about both the economic and ecologic systems allows linear mathematics to be used in analyzing both systems in a comprehensive fashion. Our linear models were reserved for the economic input-output representation. For the environmental systems, we tried to avoid this approach, and we emphasized instead the use of realistic and usually nonlinear models. Isard's study was more impact-oriented, although in a very comprehensive way. Most of the impacts were evaluated in terms of dollars for comparison to the economic benefits of the proposed projects.

In general, many of the regional science studies of the economic-ecologic interface have emphasized the input-output formulation and have used economic calculations of environmental impacts. In general, the area of environmental economics has been applied (Nijkamp 1976). Perhaps the biggest difference between our study and typical ones in the regional science field concern the emphasis on ecology and the environment. We, in a sense, tried to work from ecologic principles in order to interpret the development of society with much more emphasis on both the historic changes that have occurred in the natural world and what may take place with further development. Although we recognize economic cost-benefit analysis as a useful tool, it should be remembered that the values, in money, attributed to harvestable products and recreation from ecosystems are not a complete assessment of nature's contribution. In addition, value estimates based on money fluctuate. For example, if recreation demand falls, then an ecosystem that has been assigned recreational amenities suddenly loses value. To us, this neglects the survival value of a well-functioning environment. It still remains to be seen whether energy evaluations will give more appropriate answers than economic cost-benefit analysis for regional planning.

Adaptive Environmental Assessment and Management

Although not strictly concerned with regional analysis per se, Holling (1978) has developed an approach to environmental assessment that he calls "adaptive environmental assessment and management." This approach is intended to "integrate environmental with economic and social understanding at the very beginning of the design process, in a sequence of steps during the design phase and after implementation." According to this view, policymakers should be involved in an evaluation from the very beginning. In his book, Holling (1978) not only elaborates some general procedures and concepts, he presents several case studies ranging from forest and fishery management to regional development. The emphasis is certainly not on energy as it was in our project; however, energy development could be one example in which adaptive assessment could be used. The case studies presented are related to specific resource management issues, which is in contrast to our more general study of a total region. However, Holling's emphasis on systems analysis and simulation for dealing with the complex interactions of environment, economics, and society associated with resource development is philosophically related to our framework.

Two studies referred to by Holling are regional in nature. The first concerns development in the high mountain region area of Obergurgl, Austria. Development in this region is tightly connected to tourism, with a pronounced seasonal emphasis. Potential summer and winter demands had to be predicted and weighed against each other. There were a number of feedback effects to consider. For example, increasing numbers of winter visitors causes a high demand for housing. However, as houses are built to satisfy demand, it results in a decline in habitat diversity, which makes the resort less attractive for summer tourists. Demand predictions also had to include an assessment of available facilities to determine whether demand would be realized. Other activities that had to be considered besides tourism were farming, hotel construction, and services. Population growth, including immigration and emigration, were also important for assessing the development of the area. The interaction among tourism and other economic activities, and their impacts on the environment and land use, were all considered through a systems analysis approach.

The second regional study presented by Holling dealt with regional development in Venezuela. A river basin was studied by means of a spatial model that included precipitation, vegetation, soil, and river components. Important activities in the basin included forestry, agriculture, and hydroelectric production. Interactions were considered among the different uses of the basin. Exploitation of the land led to benefits in forestry and agriculture. However, it also led to erosion and changes in the hydrology, affecting both hydroelectric costs and outputs. A cost-benefit analysis was conducted for various development schemes. Total net benefits were evaluated over different time horizons. It was found that as local exploitation increased from no exploitation, net benefits were enhanced; however, beyond a certain level of environmental degradation (intervention), benefits declined. Thus, one could regard it as an optimum level of development. Both of these regional examples include fairly specific issues for evaluation that emphasize the systems approach as a useful tool. This is certainly in concurrence with a number of the goals of the Gotland study.

The Delaware Estuary Management Model

The study of the Delaware estuary focused on the management of the water body, with the goal of meeting water quality standards. Its importance to our studies lies in the attempt to link an ecosystem model with an economic model for the purposes of management (Kelly and Spofford 1977). The ecosystem of the Delaware estuary was represented by a system of differential equations that represented the rate of change of biotic and abiotic components. The estuary was broken up into 22 reaches; each one was represented in the model. Each reach received inputs from the next upstream reach and from municipal and industrial outputs; it discharged to the next downstream reach. The ecosystem model was then used to predict impacts along the estuary as a function of inputs from the economy. The management model consisted of an optimization model. Its objective function was to minimize total regional costs of waste water management. This determines the wastewater discharge levels, which must also satisfy ambient standards. Nonlinear objective functions were also considered to contain penalty functions for violating ambient standards.

This study had very different objectives than our own. Its strict focus on water quality concerns made it more limited in scope. It did not generally consider other resource use in relation to economic activity, such as the harvesting of the river ecosystems. Their approach went beyond ours in the use of nonlinear programming and the inclusion of spatial models; therefore, it is of general interest for our study. The design of the Delaware study was certainly in agreement with our basic approach of using systems analysis for considering ecologic and economic systems within a region.

Energy Self-Sufficiency for Hawaii

An energy study of Hawaii, as reported by Shupe (1982), is similar to our study in some ways; both concern an island system that is dependent on cheap and abundant petroleum. Although Hawaii is more distant from a mainland, both can be considered to be isolated regions. There is an increasing need to conserve energy and to develop indigenous energy sources. As presented by Shupe, the energy study of Hawaii gave a much more limited view than the systems-oriented study of Gotland. For Hawaii, solar energy technologies appear cost-effective and can provide protection from both long- and short-term global fluctuations in energy supplies and prices. Estimates were made of possible contributions from bagasse, flat-plate solar collectors, photovoltaic systems, ocean thermal energy conversion, biomass (sugarcane and tree farms), and wind and geothermal energy sources. All of them look good for supplying significant energy to meet demand. In the Gotland study, we also made estimates of potential contributions from renewable energy technologies. Our analyses were less detailed from a technologic viewpoint, but they went much further by including optimization models for the total energy supply systems. Obviously, our project as a whole was a more comprehensive energy study, with its considerations of the natural environment and the relationship of energy to economic activity. The Hawaii study is a good example of an energy study that presents a good deal of useful information for different energy technologies, but does not consider how these technologies will fit together or into an environment-economy complex.

Project Organization

The Gotland project, from the outset, was viewed by the funding agencies as a pilot project for demonstrating the applicability of energy analysis and systems methodology to the study of the interaction between natural resources and society. As such, it was never really funded at very high levels; annual expenditures varied from 75,000–600,000 Swedish crowns. At times, the project was supporting two senior researchers, the equivalent of four graduate students, and several consultants who provided expertise on problem areas of interest. The project also incorporated temporary work from students visiting Sweden who were supported for several months on fellowships from various agencies. Our experience suggests that the most effective method for project coordination and movement is to have the "big picture" or large-scale regional models be the responsibility of one or at most two project leaders who are, in their own right, also successful scientists. This focus on the large-scale model, and the integration and direction of efforts by project members, must be maintained with diligence and a sense of continuity. Ideally, the project leaders would engage in their own large-scale analysis and modeling efforts, with assistants carrying on the time-consuming chores such as data collection and computer programming. The project leaders, to keep the project integrated and communication among members high, would need to be well acquainted with issues and techniques in the natural and social sciences.

Large projects can easily tend towards disorder and disarray as the number of members increases. The problem becomes one of insuring that the members of the project continue along lines that contribute to the overall project goals and directions. In addition, it is important that individual research tasks interface with each other, where appropriate. It easily happens that junior members lose contact for periods and embark on tasks of little usefulness to the whole project. This tendency is enhanced when members of the project are located in different departments, universities, and even countries. Ideally, it would be best for the entire staff to be located in the same building. Regular meetings for presentations of ongoing research and results contribute to recognizing the totality of a project. From the outset, our particular project had a systems ecology bias because of the training of the project leaders. Understandably, many of the project members were not familiar with this approach; discipline-oriented languages, terminology, and concepts had to be bridged, especially among ecologists and economists. One way that communication could be facilitated would be to conduct a course for the first months of a project in which central concepts, ideas, and methods are taught and discussed in a classroom situation; this would lead to a greater commonality among members and facilitate communication and criticism of ideas. This approach, however, is jeopardized by a transient population on a project. It is imperative that members who are leaving the project fully document their work, especially mathematic models, so that replacements can proceed with the research. One possible means is to require frequent and detailed progress reports, but this also tends to slow down research.

From the very beginning, the Gotland project was overseen by a reference group from universities and other organizations as well as by persons particularly knowledgeable about Gotland. This is an arrangement common to Swedish research projects; it was particularly evident with regard to the Gotland project because of its interdisciplinary character and new and controversial ideas associated with the energy approach of Odum (1973). Regular meetings of the reference group were held. This served as an incentive to regularly update progress and results on the project in clear and concise presentations. Upon presentation of material, the members of the reference group would voice their criticisms. They would also make suggestions for future directions of research. There were great benefits through this process of exchange and criticism, but only after we as project leaders learned to distill the points made that we considered to be resonant with our own ideas of the project. It was common for members of the reference group to make contradictory suggestions at different meetings, because they were not involved in the day-to-day evolution of theoretic ideas or project results. With time, however, a good working relationship evolved. The reference group concept is probably appropriate for many projects, although there is a tendency, because of the group interaction process, to eliminate controversial directions of research.

Problems of Data Collection

Any scientific undertaking requires measurement and data collection, which can be very costly and time-consuming. Project goals of the Gotland study constantly encountered difficulty not only because of the nonexistence of desired data, but also because of limited funding for fieldwork. For energy analysis, data were good for total production, imports, and consumption in the industrial sector—at least for the 1970s. Energy use on a firm-by-firm basis in such important activities as agriculture and forestry, and in the urban-suburban sectors, were more problematic; much time and effort were required for making estimates of many of the subsector activities. Time series data for these activities were very sparse, so that reliable judgments on trends were difficult. Very little disaggregation of the urban sectors was done, but this certainly should be a part of any such regional study; the costs of survey work, however, would be significant. Because the resources for conducting extensive field measurements were not available, energy flows associated with the natural systems had to be estimated based on a variety of techniques. Productivity data from similar ecologic sites available in the literature were used, as well as estimates based on calculations of evapotranspiration (Eagleson 1970, Lieth 1975). Not only was it difficult to quantify the energy flows within the ecosystems, but also to estimate the harvest of energy due to human use; thus, for example, the use of firewood is an important piece of information, since it can become a significant source of heating fuel as other energy costs rise. Nevertheless, because of the extensive ecologic literature that does exist, estimates of energy flows for the natural systems probably were reasonably accurate (± 20%) and sufficient for our purposes.

Even if regional environmental statistics are not well developed in Sweden, economic data (for many purposes) were perhaps the easiest to obtain. This is not surprising, considering the importance of economics to modern societies and the resources devoted to the accumulation of statistics. Data for industrial sectors were available in the general statistics—value added, wages, man-hours, sales, and costs of raw materials. Statistics were also available on wages for the military sector, for government transfer payments, and for taxes paid to the central government. Income generated due to tourist activities had to be estimated, and (on a year-by-year basis) gross regional product (GRP) and total income could be estimated. No estimates were made concerning the "underground economy" that may operate in the region. The formulation of systems models, such as input-output, were hindered by the lack of data on flows among subsectors. Data were available only for 1975; by the time the models were debugged, they represented 7- or 8-year-old data. Trends in changes in technology, which would adjust input-output coefficients, were only analyzed for selected processes, and more effort would have been beneficial. Export and import data were more problematic, but they are available in physical measures in the harbor statistics. These were collected for the year selected for the input-output analysis. Since data at the sector level were being used, and there were about 20 sectors, any questions concerning a finer level of detail were problematic. Because Gotland's economy is dominated by small firms, adherence to confidentiality prevented the determination of firm-by-firm exchanges and, thus, the precise locational measures of economic activity.

Aside from economic measures, other data and assessment of the human system would have required a much greater investment of research time and money. Swedish population statistics are excellent; since Gotland coincided with municipality ("kommun"), accurate measures of migration, births, deaths, and age structure were easily available. The development of a more elaborate human ecology study would have required substantial interviewing and field work - an endeavor that still remains to be undertaken. In particular, a few topics that come to mind are the social ramifications of energy shortages leading to economic dislocations, the effect of resource depletion, the impact of urbanization and displaced populations, and the interaction of social organization and ecologic change.

Environmental and resource measures were probably the most difficult to accurately evaluate, especially over a number of years. There certainly was not an existing set of environmental data that could compare to economic, energy, or population statistics, although there were substantial amounts of ecologic data available regarding the Baltic coastal system. We also had to exert considerable time and effort in gathering the statistics that existed concerning water use and pollutant emissions. In addition, there were measurements of nitrates in surface water and groundwater, which allowed the construction of spatial maps of nitrate concentration. However, an extensive analysis of water quantity and quality, and the verification of associated models, would require much more extensive fieldwork. We had to limit our field investigations to one specific watershed (Lummelunda) where enough detailed data on land use and hydrology could be compiled to facilitate calibration of the water quality model. For Gotland, fieldwork would also be desirable for measuring rates of soil erosion and for evaluating the impact of management practices on agricultural and forest ecosystems. It would have also been desirable to include field investigations concerning prototype study sites, with regard to irrigated crops in agriculture and energy plantations. For regional studies, it is generally imperative to spatially map key environmental parameters and to monitor them over time; models for explaining observed changes, of course, are one of the goals for such analyses. The usefulness of remote sensing techniques, such as air photography and satellite pictures, were tested in our project; but they were too costly and time-consuming to achieve comprehensive information for the entire island. However, we still think it will be of importance in the future.

Perhaps one of the most insidious aspects of data availability was the degree to which it influences and even dictates model development. For example, the choice to consider a sector-by-sector input-output model was influenced by the fact that this kind of data would be obtained within a reasonable time and at an acceptable cost. Initial intentions of formulating firm-by-firm dynamic models were thwarted by the unavailability of the required data. Acquiring a complete set of data would probably be beyond the means of most regional studies. There will always be a compromise between what is desirable and what is available.

The Efficacy of the Systems Ecology Approach

Systems ecology has different meanings to different people. We thought of it as the systems study of entire ecosystems, including the economic systems of humans; thus, we were naturally led to consider the economy and its interactions with the environment. We think that this systems framework led us to conduct a successful project, in that we were committed from the beginning to the understanding of the entire region. We did not get diverted from our major study goal of the region as a whole by focusing on one particular piece of the region to the exclusion of everything else. This basic philosophy of systems ecology guided our whole endeavor.

One might consider that the field is broken up into two main areas: theory and methods. The methods served us in a more pragmatic way, because we could use them to model items of interest or concern to the planner, such as the impact of agriculture on water quality or the effects of price changes on the economic system. The theoretic aspects were much more removed, as in most disciplines, from the needs of planners. For example, one could talk about the energy quality of organisms high in the food chains, such as humans, but this would not influence decision-making; humans were already considered to be important. The concept of energy quality, although intuitively appealing, was not operationally useful—economic prices and technologic feasibility were probably the main concerns; for example, in considering the potential of aerogenerators for electrical production. In the section in Chapter 4 ''Future Design for Gotland: What Does History Tell Us'', where the concept of investment ratio was discussed, the ratio of purchased energies to natural energies was expressed in units of similar energy quality. This is an interesting idea, but boundaries are a problem; if a given area overdevelops by this criteria, it may only mean that more support is coming from outside the region. To a planner, this may be good or bad, but the important consideration is whether it is physically feasible.

Although systems ecology guided us to ask interesting questions, collect appropriate data, and keep cognizant of the whole, it was perhaps too oriented to the explanation of why; being scienctific, perhaps that is understandable. The planner needs to know how and what compromises need to be made in the institutional decision-making framework to achieve desired planning goals. The systems ecologist can supply much needed information on this process. We devote the remainder of this section to a critique of systems analysis and the approach of energetics, and we try to point out the weaknesses we encountered in their use.

Critique of Systems Analysis and Modeling

The Gotland study, as well as many other studies of geographic regions (Bennett and Chorley 1978, Wilson 1981), has emphasized the importance of a systems approach. To this end, it has been essential to consider the interaction among the various components in a region to assess the systemic effects of either endogenous or exogenous change. The quantitative paradigm for this has included mathematic systems modeling, with an emphasis on either differential equations or input-output analysis and computer simulation. Significant strides have been made in these approaches, but there are still glaring weaknesses.

Since one of the primary aims of systems modeling is to assess systemic impacts due to various kinds of change, and since change occurs over time, interest is focused on the dynamic behavior and its prediction over time. Especially for planning purposes, the allocation of resources is important for the future of a region, given an assessment of anticipated change. There are many problems, however, in trying to predict the future. Even if the inner workings and system interactions were well known and able to be mathematically described, there is still the uncertainty of external events. For the economic system, an input-output formulation requires an assessment of final demand or of imports, such as energy. This requires a prognosis of activity in the global economy—a task, as recent events have shown, of no simple proportions. For Gotland, it has been very difficult to foresee the export market for cement or to foresee that L.M. Ericsson would cut back on production and jobs to the extent it has in the early 1980s. The level of tourist activity can be disrupted by weather or, as in the summer of 1984, by ferry strikes. With ecosystems, both the prediction of environmental variables and the influence of external forces, such as climatic fluctuations and migrations of species, are problematic. For example, it is not a simple task to predict what levels of cod or herring can be expected in any given year; there are random as well as unknown influences involved. However, one can perform sensitivity analysis for different ranges or probable formulations of the external factors to generate some spectrum of system response. Complicating the process are the number of external factors that need to be included. Have all the important factors been included? How does the modeler know what factors will be of importance in the future?

At the internal level, the task of precisely defining a system is a trade-off between complexity and manageability. System components must be identified, behavioral descriptions of component interactions prescribed, mathematic models formulated, and measurements made for characterizing system flows and storages. The more realistic the description of a system, the more components and interactions required; this becomes a rapidly overwhelming task. Simplifying the model makes it more manageable, but subject to less realism (Figure 5.1). Some compromise has to be reached between complexity of the model, feasibility of its development, and usefulness for description and prediction. How does one know how the flows among components will change in the future? For an input-output model, this is the issue of how constant are the technical coefficients. If technology or import requirements change, these change. For example, technological changes on Gotland in cement production reduced energy consumption by 50%, which was a development difficult to foresee far ahead of time. As a model becomes more complex, the possible number of interactions and system possibilities explode; sensitivity studies can produce virtually any output, and the problem becomes one of interpreting an enormous number of possible behaviors. This is a good argument for working with relatively simple models. Greater confidence in prediction has generated the increasing time series data available for parameterizing a model; however, large models will require great investment in collection and reduction of data.

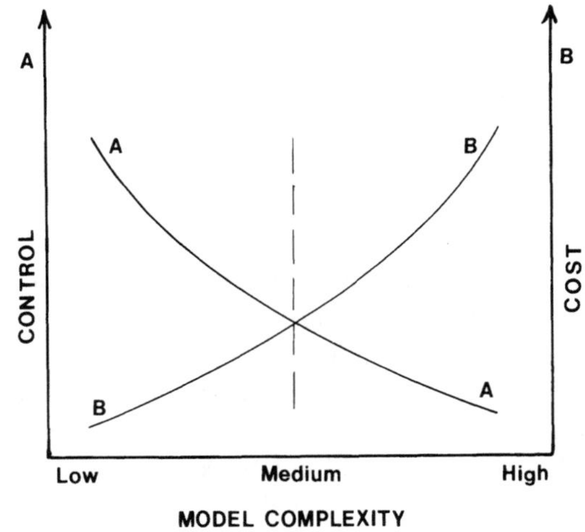

Figure 5.1. Trade-off between manageability and cost of a model as a function of complexity. There is probably an intermediate level of complexity for which control is reasonable and cost is acceptable.

The complexity vs. simplicity, micro vs. macro, or realistic vs. abstract controversies are relevant to planning and the difficulties of modeling. Relatively simple and abstract macromodels can be formulated, parameterized, and simulated; however, they may be of little use to planners who might be more interested in some specific site-related questions. Aggregation of the economy into sectors is a good example—the planner may be interested not only in the details of one firm within a sector, but in its specific location and related physical impacts as well. Certainly, an aggregated model would not supply the planner's microlevel of realism, but it might define some larger level set of constraints within which the planner would have to operate. There is a balancing act that needs to be struck between these specific microlevel needs and the more system-oriented holistic view. In some sense, the symbiosis between these two approaches occurred in the 1970s as the systems paradigm was initiated as a solution to environmental and social problems. These concepts were seen as a reaction against the reductionist approach of contemporary science.

The modeling of living systems is also made problematic by concerns other than complexity. In principle, if the fundamental physical laws are known, a system of any number of components could be modeled accurately. If there are well-known relationships between cause and effect, then complex mechanistic systems, such as machines or electromechanical systems, can be adequately modeled; it is only a matter of devoting enough resources to the task at hand. The more fundamental problem concerns the unpredictability and adaptive behavior of living systems, which become more uncertain with the high levels of intelligence and social interactions associated with humans. Although there are obvious biological and ecologic constraints on human behavior,

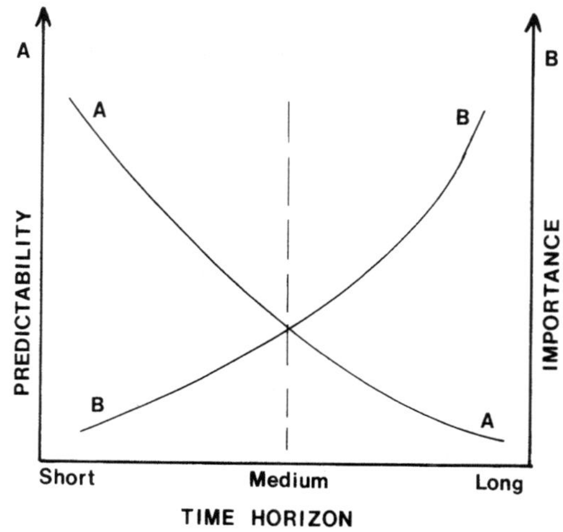

Figure 5.2. Influence of time horizon on model results. The longer the time horizon, the smaller the predictability. On the other hand, greater importance is attributed to long-time horizons because of the efficiency with which resources could be distributed.

there are also the added factors of innovation, technology, and ideas. It is difficult to predict what new ideas and technologies will emerge and, even more importantly, which ones will be selected. As further time periods in the future are considered, the uncertainty of the combined effect of a host of innovations creates additional uncertainty as to the organization of human systems; prediction becomes more difficult (Figure 5.2). For economic system models, this becomes a problem of anticipating political developments, future demands, products, activities, introduction of new firms and sectors, and changes in interactions among firms or sectors. For environmental analysis, it is a matter of delineating expected behavior of a species, of genetic change, and of the modification of the natural environment due to human action. This degree of uncertainty suggests an emphasis on scenario building, in which potential trends in the future are sketched without detailed quantitative descriptions of system components. Models, of course, can be used in this process of assessing general trends over time.

These paragraphs have stressed the weaknesses and deficiencies of modeling efforts. For all their shortcomings, however, they are an indispensable tool for dealing with complex systems. Formulation of models also helps to guide research and to determine data collection needs during the course of a project. Just the process alone, of defining a model and interpreting its dynamic behavior, helps to synthesize knowledge and generate new questions, insights, and directions for research. Even just the attempt at synthesizing the various economic, energy, and ecologic features of Gotland led to a system view of Gotland that had not previously existed. This synthesis, although not immediately implemented into the planning process, can eventually influence planners and other investigators to look at the systemic whole and to ask questions that were not asked before. The models formulated during the project were mostly of intermediate complexity—neither too abstract to be of useless generality or too detailed to be unwieldy and unmanageable (Figure 5.1). This compromise between simplicity and complexity of models is important for defining research goals and directions associated with any regional study. Although the paradigm of model building in the ecologic and social sciences has, in many instances, tried to imitate the predictability and exactness of mathematic models in classic physics, we may have to be reconciled to the fact that predicting behavior of complex living systems may remain elusive. Rather, the usefulness of models may lie better with the recognition of general properties that can be incorporated into a "Weltanschaung," which—by influencing culture, language, and perspectives—can be integrated into the complex process of societal decision-making.

Energy Analysis

Energy and its analysis can be viewed in two basic ways. On the one hand, it can be treated as just another resource whose distribution is of importance. At a more elevated level, an attempt can be made, based on a suitable theoretic development, to treat it as a measure of value (Hannon 1973, Odum 1983). The energy perspective served the Gotland project well in acting as a unifying thread to the activities in the human economy and the natural ecosystems. In addition to the distribution of energy to the various activities of humans and nature, aspects of importance were the change in energy flows over time, the quantitative relation of energy to economic activity, the comparison of energy use in human and natural systems, and the prognosis for future energy sources and costs. The combined results of such analyses, as well as the use of models, generated a comprehensive view of the structure and function of the region. The use of energy as a measure of value is still very theoretic and is subject to hypotheses and tests. For example, the notion that regions having high investments of embodied energy relative to the natural energy base would be less competitive than regions having low ratios is still to be tested across many regions; however, this would require a large data-gathering effort.

What is the prognosis for energy analysis in the future? During the 1970s the volatility of energy prices and the uncertainty of supplies generated a heightened interest in it as a critical resource. With the recognition of its constraining role for the foreseeable future, it remains a subject of interest, especially with respect to its management. However, as society adjusts to this situation, other issues will probably come foreward; for example, economic growth, unemployment, and resource management or toxic wastes. One barrier that was difficult to overcome was explaining the usefulness of energy information to planners. For the decision-maker, the problem is one of assessing information of different types in some comparative way, i.e., comparing joules to well-recognized, values such as numbers of jobs or health records, aesthetics, or money. The energy theory of value would presumably solve this difficulty by allowing all factors to be evaluated in equivalent units, but economic evaluations will probably continue to dominate. For example, in agriculture, one needs to decide what crops should be grown and the technologies to be implemented in their husbandry. Two different crops may require the same energy inputs and, in fact, produce similar energy outputs; however, one may be in demand and one not, and they may have very different nutritional attributes. The energy analysis can never become the only important criteria for consideration. Considering the externalities associated with agriculture, such as declining water quality, it is important to determine the level of nitrates, for example, in the drinking water that are acceptable for the health of the local population. Energy measures do not help much in this determination, but they may be used for computing costs for different management procedures.

Perhaps one of the more fundamental issues is the degree to which quantitative measures, especially those reduced to one common denominator, should dictate the organization and goals of a society. One reason that economics was seen as being limited in dealing with environmental issues was its inability to incorporate ecologic imperatives into its framework. Reducing systems ecology, which is the study of economic and ecologic systems, to measures of energy implies the same weakness as reducing everything to monetary units. Although energy and economic analyses enhance the understanding of complex human systems, the organizational, political, and social responses are not simplified (if at all possible) to an easily quantifiable calculus. Whether individuals are really maximizing "utility-" or systems-maximizing power remains an interesting question for future empiric work.

Institutional Interfaces

From the outset, the Gotland project was organized as a rather theoretic and academic undertaking without the funds to produce results of direct use for the solution of existing problems on the island. Emphasis was placed on the development of systematic approaches to the integration of the human economy, with the ecologic systems and the use of energy as a unifying thread. As the project progressed and various goals were met, there arose more interest by reviewers in the project as to the usefulness of this holistic view. How could it be used, especially by the planning authorities? This question was, to some extent, influenced by the tightening economic situation of the late 1970s which increasingly constrained research budgets and favored a "research for applied needs" perspective. During the latter years of the project, the shifting in focus onto the site-specific area of Lummelunda was a manifestation of the critique that management problems more germane to the interests of the planning authorities should have priority.

If interest lies in pursuing regional studies and models that interface with the institutional decision-making and are of use to planning authorities, then several points should be considered. The information should be framed in a language that is consonant with decision-makers. Theoretically oriented projects will probably have little impact unless the theory can be illustrated with very well-defined, real world examples. More pragmatically oriented studies are more easily interfaced, especially if a consensus concerning problems of interest to the decision-maker is already arrived at in the planning stages. In many instances, however, this process will obviate the inclusion of new methods and views, such as theoretic and holistic modeling efforts. A balance has to be achieved between pragmatic and theoretic endeavors. If the models, and their results, are actually going to be used to assist in decisions, then the model structure should be compatible with the institutional framework; i.e., the models should be built at a scale that is relevant to the planning problems and existing agencies. For example, it would probably be difficult to arouse interest in a hydrology model of an entire watershed if there is no agency responsible for the management of that entire watershed. This is also true for optimization models of a region as a whole. The models give solutions and indicate levels of output of different sectors to achieve an optimum. Given this information, who is responsible for making a given sector conform to the model output? In a democratic society with a free market base, this becomes problematic. Convincing arguments and education may help to change existing policies to some extent; however, to achieve optimization goals, there has to be some kind of vehicle for centralized planning at the regional level by either legislative or economic controls. This in no way implies that we advocate centralized planning, since we are only too aware of the limitations of models and the abilities of a small group of people to comprehend the complexities of society and nature. However, the insights that can be gained from studies such as ours can serve as information for the various groups and institutions involved in decisions that are ultimately made.

As was recommended for the planning authorities in Chapter 4, the most effective way to interface regional systems approaches into the planning process would probably be to incorporate a group of systems analysts directly into the planning agencies. In this manner, a group of individuals dedicated to the continuous collection of data and the formulation of systems models could interact on a day-to-day basis to address the needs of planners as well as to influence them to view the regional problems in an integrated and global sense. The role of this group would not be to pursue centralized planning, but rather would be in an advisory capacity, continually assisting the members of the planning agencies in exploring the systemic effects of a particular action. A consistent set of resource, economic, and environmental data, collected over a number of years within a systems framework, is an invaluable asset to the resolution of arising regional problems.

Interdisciplinary Work

From the outset, the Gotland project sought to unite different disciplines. In many ways, a project such as ours was an ideal way to train and educate the participants, as well as interested colleagues, in the problems of systems modeling and planning. Since we were not in a decision-making role, our efforts were to investigate and develop better approaches to regional assessment. At the very least, there has been a continual commitment to an interdisciplinary approach, especially with regard to the fields of economics and ecology. Undertaking a study as broad as that of a region obviously necessitates the consideration of many areas of knowledge; this is also a feature of ecology alone. Consequently, it is emphasized that interdisciplinary training, studies, and attitudes are essential for the understanding and management of the complex mosaic of humanity and nature. Perhaps the most important ingredient for a successful interdisciplinary project is the flexibility and willingness of the participants to operate on the interface between different disciplines, to note the similarities and differences among fields, and to be open to new concepts, perspectives, and languages. At the very least, the leaders of a project must encompass this viewpoint, even if they have specialists working for them. However, too many doctrinaire specialists can lead to much wasted effort devoted to maintaining a holistic perspective. Ideally, it is expedient to have a number of graduate students with interdisciplinary backgrounds participating in a project. Unfortunately, during times of tighter economic budgets and reductionist trends in science, it tends to become more difficult to support this kind of work in institutions of higher education. This is a discouraging development, because the problems of the future will require increasing attention to interdisciplinary matters. In our opinion, resolution of future economic-ecologic problems of society will require a commitment to interdisciplinary work—a responsibility that will be incumbent upon the systems of higher education to fulfill.

Prescription for Regional Analysis

Considering the complexities that a regional system might present, and the multifarious interests and biases that different investigators bring to a problem, it is perhaps too optimistic to expect that a general set of procedures for regional analysis can be identified. However, experience gained in the present project, as well as from other endeavors, compels us to summarize the main steps that need to be undertaken in the process of conducting a regional analysis. As a tentative outline (Figure 5.3), a number of steps are listed as follows:

1. Become familiar with the region. This especially includes personal visits and field trips to gain a first-hand impression and perspective regarding the area under study. It also means reading reports and published literature, interviewing people in positions of power as well as local residents, and gathering statistics from various sources. Organize knowledge of the human system, the ecologic systems, and the geography and better define what the boundaries of the study region will be.
2. Based on the information gathered, identify and formalize (either conceptually or mathematically) the activities and interactions that occur in the

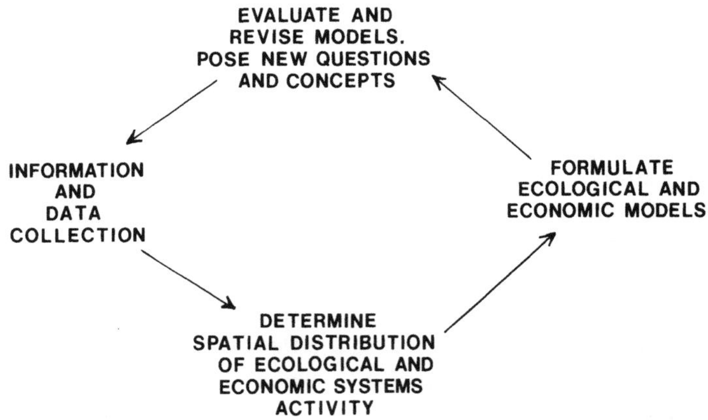

Figure 5.3. Simplified prescription for the process of doing regional analysis.

region. Identify key issues and problems that are relevant. At first, this can be an aggregated view; but then, it should be disaggregated spatially in that the location of industrial activities, human settlements, natural ecosystems, and environmental features are important to the detailed understanding of processes in the region. Rough maps of land use, economic activity, energy flows, and other resources might be made at this point, although this would depend on the quality of the information available. Tasks outlined in step 1 may also be undertaken again at this point.

3. Formalize some models, whether they are conceptual or quantitative, regarding aspects of the environmental and human systems that are anticipated to be affected by economic activity. This step requires choices about the ecosystems to be studied, the types of models to be employed, hypotheses to be tested, and issues that should gain attention. Although the familiarization of the region since step 1 may have entailed the collection of quantitative data both from statistical sources and fieldwork, the modeling process focuses attention on additional data needs and requirements. After some models are formulated—if enough information and data are available to evaluate them—depending on the results, recommendations may be made for the modification of economic activity to ameliorate anticipated undesirable impacts. New ideas and concepts may emerge, and this may lead to proposals for technologic improvements or to strategies for minimizing economic-ecologic conflicts.
4. This process is a continuing one with new data collected or measured, impacts and models evaluated, and reorganization occurrences. New solutions may be proposed to meet some defined criteria or goals for the region as a whole.

In reality, any research process cannot be illustrated simply by listing a number of steps in cook book-like fashion. In a multidisciplinary project of this kind, many people will be working simultaneously on the different steps outlined above as each person concentrates on one particular process. Serendipity is an important factor, since new ideas or events may arise spontaneously and unexpectedly, directing the whole research project in unforeseen directions. The failure of verifying hypotheses or the emergence of data that indicate unexpected problems makes an anticipated well-oiled project begin to encounter friction, but this is to be expected. Finally, the extent and emphasis on different aspects of the research process is dependent on the available economic and personnel resources; how large a staff can be supported, how much fieldwork conducted, or how much equipment purchased, and the extent to which complex computer models can be simulated are, of course, very much dependent on available funding and the time frame over which funds are committed.

The Issue of Carrying Capacity

As mentioned briefly in the section in this chapter concerning the study of the eastern Fiji islands, the notion of carrying capacity is relatively easy to deal with for simple subsistence economies, as long as reasonably accurate estimates of the biological potential of indigenous ecosystems can be made. When society becomes more complex and an economy becomes open to trade in energy and goods, then the ecologic productivity is only one factor contributing to the support of the population. If fuels can be brought in from the oustide, goods can be produced for export to generate payment for imports. One can argue, then, that the total work done by a given economy would be related to the sum of indigenous ecologic productivity and imports of fuels. Assuming the indigenous energies constant, and no changes in technology or efficiency, a reduction in external sources of energy would either support the same or a greater population at a lower standard of living or, conceivably, a lower population at the original or higher standard of living. Presumably, therefore, carrying capacity could be related to the amount of work done. However, it is more complicated, because it depends on the type and organization of the activities engaged. Employing external energies in the transformation of forests, as happened on Gotland in the nineteenth century, could lead to a diminution in the entire carrying capacity of the island due to excessive soil erosion and declines in water quality and quantity. Other activities, such as an energy-intensive cement industry exploiting limestone on a small part of the island, could enhance carrying capacity. Two different activities may entail similar consumption of energies, but may produce very different economic benefits and environmental disturbances.

The time frame considered is critical. For example, it may be temporarily possible to enhance the carrying capacity of a region by increasing agricultural productivity through drainage and fertilizers. After several decades, however, agricultural productivity may be diminished because of a loss of organic soils and water supplies. This would lower the carrying capacity until the soil and water resources were restored, which could take hundreds of years. Other resources could be exploited for export in order to import needed goods and enhance the carrying capacity.

Energy is obviously not the only concern for carrying capacity limits. Gotland has plentiful supplies of energy, but its population will still be limited by supplies of water of acceptable quality. Potentially, if energy was extremely plentiful and cheap to import, the desalinization of seawater might remove the water constraint on growth. This could be related to the notion that increased energy investment above and beyond the indigenous natural productivity will enhance the carrying capacity of the island. Increasing the supply of water through energy expended in desalinization is work done in addition to the production of fresh water by the natural hydrologic cycle. Again, how the energy is expended is crucial to the resulting impact. Therefore, the open nature of regions makes it difficult to define a meaningful intrinsic carrying capacity.

The whole issue of carrying capacity, although at a global scale, has been a central part of the "limits to growth" debate being carried on since the first global model was published by Meadows et al. (1972). Will resource scarcities set limits or will new discoveries of resources, plus technologic breakthroughs, enhance the size of human populations that can be supported? This would imply a very high-density living in urban agglomerations, with a consequent modification of the landscape to one that is intensely managed. Such questions are also relevant at the regional level. Although food production on Gotland could support about six times the resident population, the existence of this many people would have a great impact on the island (assuming there was enough water to meet their demands). If it were not possible to attract economically competitive industries to employ this larger population, then the average standard of living would be much lower. Therefore, the important consideration is that of the number of people vs. quality of life. It is not easy to establish a definite rule with regard to carrying capacity because of the mix of resource, economic, environmental, social, and cultural factors that interact.

During the final weeks of revision of this manuscript, the terrible events associated with famine in Ethiopia were widely publicized. Although the reasons are complex, the famine was partially due to a population exceeding the resource limits of the landscape. Although our study has not had to confront the risks and benefits associated with economic development and the potential of such large environmental disasters, this incident underscores the importance of planning that encompasses economic and environmental trade-offs. Different areas and regions will have different environmental issues to confront. In the case of Gotland, it was not necessary to address problems associated with rapid population growth or high rates of industrial development; the society has more time to generate solutions to perceived or anticipated problems. For many regions and countries, however, the scale and rate of change of development is much vaster and the potential consequences (as in Ethiopia) are much more serious. This makes the need for a systems view of economies and the environment even more relevant. Economic development must broaden its perspective to include considerations of natural resource conservation and management of ecologic resources to provide for the long-term viability of economies and their environments. Developing regions can learn from the mistakes of the developed world. In trying to achieve a viable balance between development and environmental protection, the systems ecology approach can serve as an important template for the incorporation of the ecologic perspective into the traditional view of economic development.

Appendix I. Notes to Selected Figures

Notes to Figures 2.4 and 2.5

Documentation of land-use data.

1. Analyses of the 1700 A.D. taxation map of Gotland give the following distribution of land use (Lindquist, personal communication):

Forests	1786 km^2	(Supposed to include coniferous forests and heath with trees)
Arable land	195 km^2	(One-third to one-half of this area probably fallow)
Meadows	410 km^2	(Supposed to include beaches, meadows, and heath without trees)
Deciduous pastures	448 km^2	(Supposed to include almost all deciduous woods)
Mires	236 km^2	
Lakes	47 km^2	
Others	7 km^2	(Beaches and urban areas).

2. According to Olofsson (1945), the area of arable land was 150 km^2 in 1805. Other land use except forests and heaths (see note 8) are assumed to be the same as in 1700.

3. Munthe (1913) estimated the land use in 1841 to be:

Forests	2276 km^2	(Supposed to include deciduous and coniferous forests, pasture, and heath with trees)
Arable land	176 km^2	
Meadows	352 km^2	
Other	316 km^2	(Mires, lakes, beaches, heaths, and urban areas).

4. Area of arable land, 1805–1937, from Olofsson (1945); 1937–1980, from the Swedish Statistical Bureau.

5. According to Sylwan (1895), 100 km^2 of wetlands (mires and lakes) had been drained in 1890.

6. Area of natural meadows, 1850–1918, from Olofsson (1945). Area of natural meadows, 1920–1980, assumed to be equal to the difference between the existing natural meadows in 1841 (= 352 km^2) and the total area of cultivated leyland in different years.

7. Area of deciduous forests supposed to have decreased in proportion to the increase of arable land minus the area derived from drained wetlands and cultivation of natural meadows.

8. Coniferous forests and heath systems. In 1898, the total area of forests on Gotland was 1381 km^2. The area of heath was 525 km^2, of which 396 was forested and 129 km^2 was only sparsely vegetated (Hesselman 1908). In 1950, the total area of forests was 1350 km^2 (Linné personal communication).

   In several historic documents, the decrease of productive forests on Gotland during the eighteenth and nineteenth centuries has been pointed out. This has been indicated in the diagram, together with a corresponding increase in the area of heath.

9. Present land use on Gotland summarized by Jansson and Zucchetto (1978a).

Notes to Figure 2.23

Commercial fertilizer application: 6308 • 10^3 kg nitrogen (N).
Based on fertilizer import statistics (see below).

Nitrogen fixation:
30 kg N ha^{-1} yr^{-1} for unfertilized pasture • 8000 ha = 240 • 10^3 kg N
1 kg N ha^{-1} yr^{-1} for other cropland • 82,400 ha = 82 • 10^3 kg N
2 kg N ha^{-1} yr^{-1} in forests • 150,000 ha = 300 • 10^3 kg N
other vegetation (heath, and so on) 1000 • 10^3 kg N.

Denitrification:
1 kg N ha^{-1} yr^{-1} in cropland • 82,400 ha = 82 • 10^3 kg N
0.91 kg N ha^{-1} yr^{-1} in forests • 150,000 ha = 137 • 10^3 kg N.

Precipitation:
11 kg N ha^{-1} yr^{-1} on cropland • 82,400 ha = 906 • 10^3 kg N
6.6 kg N ha^{-1} yr^{-1} in forests (Viro 1953) • 150,000 ha = 990 • 10^3 kg N.

Runoff and leaching:
Cropland:
30 kg N ha^{-1} yr^{-1} for sandy soil • 12,400 ha = 372 • 10^3 kg N
30 kg N ha^{-1} yr^{-1} for peat soil • 15,000 ha = 450 • 10^3 kg N
20 kg N ha^{-1} yr^{-1} for glacial till • 51,200 ha = 1,024 • 10^3 kg N.
Forests:
1.5 kg N ha^{-1} yr^{-1} for glacial till • 119,200 ha = 179 • 10^3 kg N
2 kg N ha^{-1} yr^{-1} for sandy soils • 2000 ha = 40 • 10^3 kg N.

Harvest in agriculture (see below).

Harvest export from agriculture 619 • 10^3 kg N.

Animal consumption:

Fodder imports	186 • 10^3 kg N
Local fodder, totally	1833 • 10^3 kg N including
Tops of sugar beets	28 • 10^3 kg N
Grain	126 • 10^3 kg N
Ley	1257 • 10^3 kg N
Green forage	37 • 10^3 kg N
Straw	8 • 10^3 kg N
Hay	12 • 10^3 kg N
Other	148 • 10^3 kg N
Grazing, assuming 50–70% of available area being grazed	175-262 • 10^3 kg N.

Manure (see below).

Food imports, totally:	220 • 10^3kg N including
Grain	80 • 10^3 kg N
Other	140 • 10^3 kg N.

Sewage discharge:
12.1 g N $capita^{-1}$ day^{-1}
54,000 persons • 365 days = 238 • 10^3 kg N yr^{-1} (Karlgren 1972).

Nitrogen uptake by humans in drinking water:
2 mg N l^{-1} • 4.4 • 10^6 m^3 = 8.8 • 10^3 kg N.

Primary production in forests:

Using the following age distribution among trees: 24% < 38 years old, 18% between 38–70 years old, and 58% > 70 years old—the uptake of N was calculated as follows (N uptake by trees):

$(0.24 \bullet 19.0) + (0.18 \bullet 28.2) + (0.58 \bullet 38.8) = 32.1$ kg $ha^{-1}yr^{-1} \bullet 150{,}000$ ha $= 4815 \bullet 10^3$ kg N.

N uptake by ground vegetation:

$(0.24 \bullet 12.2) + (0.18 \bullet 10.6) + (0.58 \bullet 23.0) = 18.2$ kg ha^{-1} $yr^{-1} \bullet 150{,}000$ ha $= 2727 \bullet 10^3$ kg N,

assuming steady-state, this equals litterfall.

Litter fall: Trees $= (0.24 \bullet 3.3) + (0.18 \bullet 5.7) + (0.58 \bullet 11.0) = 8.2$ kg ha^{-1} yr^{-1} $\bullet$ 150,000 ha $= 1230 \bullet 10^3$ kg N.

Ground vegetation: (see above) $2727 \bullet 10^3$ kg N.

Harvest in forestry: 22% of net primary production $= 1053 \bullet 10^3$ kg N.

Export from forestry: $301 \bullet 10^3$ kg N.

Application of Fertilizer to Cropland on Gotland[a]

Crop	Area (km^2)	Application Rate (kg ha^{-1} yr^{-1})	Total Application (10^3 kg yr^{-1})
Pasture	80	180	1440
Winter rape	30	175	525
Ley	190	125	238
Sugar beets	40	110	440
Potatoes	10	100	100
Spring wheat	10	100	100
Winter wheat	80	100	800
Oats	30	90	270
Barley	220	85	1870
Mixed grain	10	75	75
Winter rye	60	65	390
Spring rape	10	60	60

[a] Data from County Agricultural Board.

Nitrogen in Harvest and Total N Uptake, 1972

	Harvest[a] (10³ kg)	Grain (%)	N in Grain Harvest[b] % N	tN	N in Straw % N	tN	N in Roots[c] % N	tN	Total N Uptake
Winter wheat	23,980	43	1.8	186	0.5	68	1.5	108	362
Winter rye	15,303	40	1.6	97	0.5	45	1.5	68	210
Barley	46,290	50	1.6	370	0.6	139	1.5	209	718
Oats	6,750	50	1.7	57	0.6	20	1.5	31	108
Ley	81,920	—	2.2	1802	—	—	1.5	3684	5489
Potatoes	15,400	60 (roots)	0.35	32	0.65	40	—	—	72
Sugar beets	155,903	60 (roots)	0.20	187	0.45	280	—	—	467
Winter rape	1,600	30	3.5	17	0.8	9	1.5	7	33
Spring rape	230	30	3.5	2	0.8	1.3	1.5	1.5	5
Winter turnip rape	1,180	30	3.5	12	0.8	6.6	1.5	5.4	24
Spring turnip rape	840	30	3.5	9	0.8	4.7	1.5	6.5	17
Other	130	—	1.5	2	—	—	—	0.6	3
Total	349,303			2773		614		4121	7508

[a] In the case of ley, the data for seeds is not given; for sugar beets and potatoes, root weights are given.
[b] Information from Svanberg (1972).
[c] Calculated from below/above-ground ratios of 3:1 for ley and 0:3 for other crops.
Total N in harvest: 3387 • 10³ kg N (grain + straw).
N returned to soil: (straw + roots) = 4735 • 10³ kg N.

Nitrogen (N) in Manure for 1972 Based on Stable Production During 8 Months. Total for 12 Months Estimated Based on Equal Production per Month[a]

	Number	Nitrogen per Animal (kg N yr⁻¹)	(solid + liquid)	Total Nitrogen (10³ kg N yr⁻¹)
Cows				
Milk	16,595	35.7	(33.6 + 2.1)	592
Calves	15,210	15.8	(14.8 + 0.95)	240
Heifers	17,701	35.7	(33.6 + 2.1)	632
Pigs				
Sows and piglets	7,132	15.1	(8.5 + 6.6)	108
Others	46,206	4.1	(2.3 + 1.8)	189
Sheep	84,696	3.6		305
Poultry	383,121	0.8		306
Horses	1,225	26.8		33
Total, 8 mo				2405
Total, 12 mo				3608

[a] Data from County Agricultural Board and Supra Fertilizer Ltd.

Nitrogen Budget in Forest System (All Values in kg N ha^{-1})

	Age of Stand		
	< 38 yr	38–70 yr	> 70 yr
Trees			
N uptake by Scots pine	19	28.2	38.8
Litterfall	3.3	5.7	11
% of uptake	17	20	28
Biomass storage	74.4	130	186
Above ground	53.3	108.7	150.3
Below ground	21.1	21.2	36
Ground Vegetation			
N uptake	12.2	10.6	23
Litterfall	12.2	10.6	23
Biomass storage	38.5	42.6	59.4
Above ground	28.6	27.2	43.4
Below ground	9.9	15.4	16
Total litterfall	15.5	16.3	34
N returned to soil by litter decomposition	2.8	5.3	10.4
Total litterfall (%)	18	32.6	30.6

Notes to Figure 2.24

Annual Net and Gross Metabolic Work for Photosynthetic Systems (The Values Were Approximated From the Literature and No Field Measurements Were Made on Gotland)

		Total				Intensity (m^2)			
		Net		Gross		Net		Gross	
Natural Systems	Area (km^2)	kg (10^9)	kJ (10^{12})	kg (10^9)	kJ (10^{12})	g	kJ (10^5)	g	kJ (10^3)
Coniferous forests (merchantable)[a]	1260	0.89	16.5	2	38	700	0.13	1600	30
Mixed deciduous forests[b]	209	0.19	3.5	0.43	7.95	900	0.17	2060	38
Heath with trees[c]	139	0.03	0.56	0.07	1.40	240	0.04	540	10
Grassland heaths[d]	268	0.03	0.54	0.06	1.08	110	0.02	220	4
Wooded pastures[e]	203	0.14	2.6	0.32	5.89	700	0.13	1600	29
Mires[f]	31	0.08	1.4	0.12	2.10	2500	0.46	3750	69
Lakes[g]	25	0.01	0.2	0.02	0.31	400	0.08	600	12.6
Beach systems[h]	62	0.03	0.62	0.08	0.6	540	0.1	1230	9.6
Baltic coastal systems	2449			0.88	17.6			360	7.2
Total	4646			3.97	74.88				
Agriculture	871				25.7		0.18		29.6
Total photosynthesis					100.6				

[a] See Malkönen (1974).

[b] Total above-ground net production assumed to be 6500 kg ha^{-1} dry weight above-ground production (Hytteborn 1975, Persson 1975). Assume the ratio of below-ground to above-ground production to be 40%, so that below-ground production = 2550 kg ha^{-1}. Total net production = 9050 kg ha^{-1} or $0.17 \bullet 10^9$ kJ ha^{-1} yr^{-1}. Assume gross production = Net/0.44 = 20,568 kg ha^{-1} (E.P. Odum, 1971) or $0.38 \bullet 10^9$ kJ • ha^{-1}.

[c] Productivity of heathlands assumed to be about one-third of merchantable forests (Linné personal communication), so that net production = 2359 kg ha^{-1} yr^{-1} = $0.04 \bullet 10^9$ kJ ha^{-1} yr^{-1} and gross production = 5360 kg ha^{-1} yr^{-1} = $0.1 \bullet 10^9$ kJ ha^{-1} yr^{-1}.

[d] Assume 1100 kg ha^{-1} yr^{-1} net production (Sjörs 1954); gross production is $2 \bullet 1100$ = 2200 kg ha^{-1} yr^{-1} or $0.02 \bullet 10^9$ kJ ha^{-1} yr^{-1}, respectively.

[e] The deciduous-pasture mixture is assumed to consist of patches of deciduous trees and grassland. Net productivity of deciduous = 9050 kg ha^{-1} (see note 2). Net productivity of grassland assumed to be 5000 kg ha^{-1} (Whittaker and Likens 1973). If total area is divided evenly between systems, then average net production = 7025 kg ha^{-1}. Gross production = 7025/0.44 = 15,966 kg ha^{-1} = $0.29 \bullet 10^9$ kJ ha^{-1} yr^{-1}.

[f] Assume net production of mires = 25,000 kg ha^{-1} yr^{-1} (Whittaker and Likens 1973), and the ratio of gross/net = 1.5 (Young 1975), so that gross production = 37,500 kg ha^{-1} yr^{-1}. In energy terms, these are $0.46 \bullet 10^9$ kJ ha^{-1} yr^{-1} and 0.69×10^9 kJ ha^{-1} yr^{-1}.

[g] Estimated gross production of 300 g C m^{-2} yr^{-1} and net production of 200 g C m^{-2} yr^{-1} from Graneli (1977); or in energy terms, $0.13 \bullet 10^9$ kJ ha^{-1} yr^{-1} and $0.08 \bullet 10^9$ kJ ha^{-1} yr^{-1}, respectively.

[h] Most productive land consists of beach meadows with an area of 26 km^2. Assume above-ground net productivity = 380 g $m^{-2}yr^{-1}$ (Wallentinus 1973). Below-ground percentage unknown, but assumed to be 30%. Total net production = 543 g m^{-2} yr^{-1} and gross production = 1234 g m^{-2} yr^{-1} = $0.23 \bullet 10^9$ kJ ha^{-1} yr^{-1}.

Notes to Figure 2.26

Weight, Energy Value, and Protein Value of Fish Catch for Gotland Fisheries.

	1920			1944			1973		
	Weight[a] (10^3 kg)	Energy[b] Value (10^6 kJ)	Protein[c] Value (10^6 kJ)	Weight (10^3 kg)	Energy Value (10^6 kJ)	Protein Value (10^6 kJ)	Weight (10^3 kg)	Energy Value (10^6 kJ)	Protein Value (10^6 kJ)
Herring	1418	9497	4235	1732	11,601	5172	1850	12,391	5526
Cod	488	1655	1549	204	692	648	575	1950	1825
Flounder	102	376	294	138	509	398	162	597	467
Salmon	34	306	110	87	783	282	259	2331	840
Other	49	205	126	96	402	247	316	1322	813
	2091	12,039	6314	2257	13,987	6747	3162	18,591	9471

[a] Total catch in metric tons found in fishery statistics.

[b] Energy values of fish were obtained from Abramson (1976). These were: herring = 6.66 kJ g^{-1}; cod = 3.34 kJ g^{-1}; flounder = 3.68 g^{-1}; salmon = 8.99 kJ g^{-1}. Other species were assumed to have a value of 4.2 kJ g^{-1}. The energy values calculated for the fish did not include the energy value of the internal organs.

[c] Value of protein was assumed to be 17.16 kJ g^{-1}. Weight of protein per weight of fish for different species was: herring = 0.174 g g^{-1}; cod = 0.185 g g^{-1}; flounder = 0.168 g g^{-1}; salmon = 0.189 g g^{-1}; other = 0.150 g g^{-1}.

Approximations for energy consumption for fishing come from the Rationing Unit, Gotland County Administration. The amount of diesel fuel used for salmon fishing was 349.2 $m^3 \bullet 35.6 \bullet 10^9$ J = $12.4 \bullet 10^9$ kJ. The amount of energy used for trawling was 630 m^3 diesel $\bullet$ $35.6 \bullet 10^9$ J = $22.43 \bullet 10^9$ kJ. The percentage by species for trawling was: cod = 16.9%, herring = 70.3%, flounder = 2.9%, and other species = 9.9%; the energy cost for each species was allocated by these percentages.

The total energy used for coastal fisheries was 66 m^3 gasoline plus 20.4 m^3 diesel or $66 \bullet 10^9$ J + $20.4 \bullet 35.6 \bullet 10^9$ J = $2.8 \bullet 10^9$ kJ. The percentage by species for coastal fisheries was: cod = 40%, herring = 17.5%, flounder = 23.9%, and other species = 18.6%. Energy cost was allocated by these percentages. In summary, the energy costs for each species were:

	Gasoline		Diesel		Total Energy
	(m^3)	(10^9 kJ)	(m^3)	(10^9 kJ)	(10^9 kJ)
Salmon	—	—	349.2	12.4	12.4
Cod	26.4	0.83	114.7	4.08	4.91
Herring	11.6	0.36	446.6	15.9	16.26
Flounder	15.8	0.5	23.2	0.83	1.34
Other	12.3	0.39	66.2	2.36	2.72
Totals	66.1	2.08	999.9	35.57	37.63

Appendix II. Notes to Selected Tables

Notes to Table 2.2

Annual work is the gross photosynthesis and is a measure of the total work performed by plants.

Terrestrial systems consist of forest, heathland, pasture, marsh, and beach systems. See the section on "Agricultural Systems."

The Baltic coastal system is contiguous to the coast of Gotland and includes that part of the Baltic Sea out to the 30 m depth contour.

A profile of average wind velocity was assumed. To approximate the kinetic energy within a 100-m height, a linear increase of velocity from 4–8.2 m sec^{-1} was assumed from a 0–50 m height and a constant velocity of 8.2 m sec^{-1} from 50–100 m height. The length of the island from north to south was approximated as 140 km; the volume of air passing per year was $3.15 \cdot 10^{15}$ m^3 yr^{-1}. Kinetic energy = ½ (0.0013 g • cm^{-3}) ($3.15 \cdot 10^{21}$ cm^3 yr^{-1}) (715 cm sec^{-1})^2 = $105 \cdot 10^{12}$ kJ yr^{-1}.

This accounts for the lateral transfer of energy within 100 m of the ground from outside the island. There is also turbulent diffusion of energy from above the island down to the ground. This is difficult to calculate, but if an eddy diffusion coefficient of $K = 10^4$ cm^2 sec^{-1} is assumed along with a gradient of about 0.08 m sec^{-1} m^{-1} of height, then this energy transfer is approximately:

$$\tfrac{1}{2}\,\rho\,\frac{v^2K}{h} = \tfrac{1}{2}\,\frac{(0.0013\ \text{g cm}^{-3})(6.2\ \text{m sec}^{-1})^2\,(10^4\ \text{cm}^2\ \text{sec}^{-1})}{50\ \text{m}}$$
$$= 1.52 \cdot 10^4\ \text{kJ m}^{-2}\ \text{yr}^{-1}.$$

For the total island this wind energy is $47.4 \cdot 10^{12}$ kJ yr^{-1}. The total wind energy is thus $152 \cdot 10^{12}$ kJ yr^{-1}. On a per area basis for the island, this is $4.9 \cdot 10^{4}$ kJ m^{-2} yr^{-1}.

On the other hand, if we consider the cross-sectional area of systems through which the wind is blowing, we get an energy intensity of $10.9 \cdot 10^{6}$ kJ m^{-2} yr^{1}.

An average power over the year for waves in the Gotland region is about 4 kW/m of wavefront. There is about 485 km of coastline, so that the total energy delivered is $56 \cdot 10^{12}$ kJ yr^{-1} (Anonymous 1977a).

If we assume the waves break within 100 m of the shore, this results in an energy intensity of $1.15 \cdot 10^{6}$ kJ m^{-2} yr^{-1}, and this energy is doing work in the nearshore area and on the beach system.

The chemical potential of rainwater with respect to the Baltic Sea is:

$$\Delta F = \Delta F_O + nRT \, 1n \frac{C_2}{C_1}$$

$$\Delta F = 0 + \frac{1}{35}(1.99)(300) \, 1n \frac{1.2}{7000}$$

$$\Delta F = 147.9 \text{ cal g}^{-1} \text{ solute}$$

where C_2 = concentration of solute in rainwater = 1.2 ppm.
C_1 = concentration of solute in Baltic Sea = 7000 ppm.
ΔF = -147.9 cal g^{-1} solute • 7 g solute 1^{-1}.
ΔF = 1.04 kcal 1^{-1} = 4.35 kJ 1^{-1}.

Rainfall (1972) = $1.4 \cdot 10^{9}$ m^{3} = $1.4 \cdot 10^{12}$ 1 yr^{-1}.

Total free energy of mixing = $4.35 \cdot 1.4 \cdot 10^{12} = 6.1 \cdot 10^{12}$ kJ yr^{-1}. On an area basis for the island, this amounts to $1.96 \cdot 10^{3}$ kJ m^{-2} yr^{-1}.

The average elevation for Gotland was approximated by "eyeballing" topographic maps of the island to arrive at an average elevation of about 13 m above sea level. The gravitational potential energy (PE) of water working on the environment is thus:

PE = (1 g cm^{-3}) (1.4 10^{9} m^{3}) 10^{6} cm^{3} m^{-3} 980 cm sec^{-2} 1300 cm
PE = $1.8 \cdot 10^{11}$ kJ yr^{-1}. On an area basis for the island, this amounts to 58 kJ m^{-2} yr^{-1}.

Notes to Table 2.13

Area contributing to the agricultural model consisted of arable land, grazed managed land, and an estimate of the forest area used for grazing, which was obtained from a study of 22 sheep farms. The total area amounted to 87,097 ha with a total solar insolation of $87{,}097 \cdot 10^{4}$ m^{2} (10,184 kJ m^{-2} day^{-1}) (365 days) = $3.24 \cdot 10^{15}$ kJ yr^{-1}.

Data on energy use for agriculture made by the Rationing Unit, Gotland County Administration. The conversion figures to energy include the energy for extraction, transport, and so on in accordance with Anonymous (1975). (A factor of 1.21 is used to account for production, and so on, in correcting the energy content.) Gasoline = $8 \bullet 10^9$ kJ. Diesel + fuel oil = $240 \bullet 10^9$ kJ. Kerosene = $21.3 \bullet 10^9$ kJ.

From Anonymous (1945–72): Corrected for the fact that 55% of the electricity to farms is used in households; $51.9 \bullet 10^9$ kJ electrical energy is thus used in production and $62.8 \bullet 10^9$ kJ is consumed by agricultural households.

The cost of pesticides based on Swedish averages = $20.9 \bullet 10^9$ kJ. Energy needed for seed production is the actual energy needed to produce the seed—not the energy content of the seed = $4.6 \bullet 10^9$ kJ. Cost for veterinarian services, transports, and so on = $77 \bullet 10^9$ kJ. Total = $102.5 \bullet 10^9$ kJ.

Energy costs for buildings is based on the actual construction on Gotland in 1972 for farm purposes. The energy figure is from Anonymous (1975) and includes both the energy content of materials, production of them, and construction work excluding labor. Total = $23.9 \bullet 10^9$ kJ.

The cost for drainage is based on the actual area under drainage with covered drains. Amount of pipe and energy values for production of pipes and laying are the average Swedish values. Write-off period, 40 years. The value of yearly replacement; total = $15 \bullet 10^9$ kJ.

Cost for irrigation is based on actual area under irrigation and average Swedish values for production of irrigation pipes, pumps, and so on. The value is the replacement of equipment per year. The cost for irrigation = $0.4 \bullet 10^9$ kJ.

The cost for tractors is based on the actual numbers of tractors, average real life of tractors (15 years), and data on production and maintenance from Anonymous (1975). The basic data are from Berry and Fels (1973). Electric energy is thus included according to U.S. energy analysis convention. The cost for tractors = $113 \bullet 10^9$ kJ.

Machinery data from Anonymous (1975) and County Agricultural Board have been checked against each other. The figure given is based on the average machinery needed for the production attained on Gotland. The alternative way of using input-output data is feasible, since there seems to be no way of separating out the value of machinery bought during 1972 from other goods. The actual number of combine harvesters on Gotland is available; however, using this figure as a norm is probably less satisfactory, since a large proportion seem to be old and replacement will be with more energy-intensive machinery. Combine harvesters may not be a good index of mechanization anyway. The figure is probably one of the more important inaccuracies in the material. The cost of machinery = $172 \bullet 10^9$ kJ.

The value is for the production and maintenance of the cars used in agriculture calculated in accordance with Berry and Fels (1973) on the basis of mileage. Total = $2.1 \bullet 10^9$ kJ.

Energy cost for fertilizer based on a value of 19.2 kWh kg^{-1} for pure N, 2.78 kWh kg^{-1} P, and 0.63 kWh kg^{-1} K, and the actual amount imported. The fertilizer use seems to be excessive in relation to the harvest. A rough comparison to average Swedish data indicates an excess level of fertilization equivalent to about 20–25 kg/ha of N. This may be accounted for if it is noted that the total arable land on Gotland is about 3% of the arable land in Sweden, but Gotland produces slightly less than 10% of the total amount of sugar beets. This might account for some of the discrepancy when calculating averages.

$$19.2 \cdot 6{,}319{,}000 \text{ kg} + 2.78 \cdot 1{,}916{,}000 \text{ kg} + 0.63 \cdot 47{,}000{,}000 \text{ kg} = 129.28 \cdot 10^6 \text{ kWh} = 465 \cdot 19^9 \text{ kJ}.$$

Energy content and energy cost for imported grain is obtained from official import statistics. Feed grain consists of 80% barley and 20% oats. Energy content = $71.2 \cdot 10^9$ kJ, and energy cost = $17.6 \cdot 10^9$ kJ.

Data on imported feed is based on actual import data and an average energy content of feed of 4.9 kWh $\cdot$ kg^{-1}. The energy needed to produce feed is not included. Based on the average data for producing (growing and processing, and transporting) soy cake, it should be about $71 \cdot 10^9$ kJ.

Biomass production data is based on the rough calculation that below-ground parts are equivalent to above-ground parts for grains, and three times the above-ground for ley at time of harvest. The figure for biological harvest has been corrected for this. Obviously, the accuracy of the figure is low. Gross = net/0.6 = 15,446/0.6 = $25{,}743 \cdot 10^9$ kJ (E.P. Odum 1971). Maintenance = gross − net = $10{,}297 \cdot 10^9$ kJ.

Biological harvest is the total harvest plus straw and other above-ground parts. On the average, straw is equivalent to 90% of the energy content of grains and the rest is 23% of the energy content. The actual harvest amounted to $3474 \cdot 10^9$ kJ based on the Central Bureau of Statistics. The accuracy of these statistics is generally considered to be good. Conversion values are from Anonymous (1975). Total biological harvest = $6735 \cdot 10^9$ kJ.

The amounts of organic matter returned to the soil is straw plus other above-ground parts − the amount of straw and sugar beet tops used as animal feed. The estimate is likely to be somewhat too high, since some straw may be burned and some straw is used as bedding. Of the latter, a large proportion returns to the soil with the manure. Total amount returned to soil = $1850 \cdot 10^9$ kJ.

The amounts used in animal production are based on data from the County Agricultural Board. Some of this comes from grazing. This is why this value ($3646 \cdot 10^9$ kJ) is higher than the harvest figure of $3474 \cdot 10^9$ kJ.

Export data is actual data from statistics. The rest is residual.

There are estimates of average Swedish values of grain, milk, and potatoes used for food. If one assumes that 3% of the meat goes to farm households, based on the values used for taxation purposes, one gets a value of $14.2 \cdot 10^9$ kJ for farm households. This is based on percentages of the production. Since a large proportion of the population on Gotland is employed in agriculture, the figure is probably too low. The alternative method would be to calculate how many kJ the average person takes out of the farm yields. Using the figure $1850 \cdot 10^3$ kJ yr^{-1} and per employee in farming, and the number of employees in agriculture on Gotland in 1970, this gives $9.6 \cdot 10^9$ kJ. The choice of the higher value gives a conservative estimate as to what is left over.

Edible animal products ($430 \cdot 10^9$ kJ) are based on data from the County Agricultural Board, Arla and Farmek. Conversion factors are from Anonymous 1975.

Other animal production (blood, bone, and so on) is calculated as the same percentage of the energy value for edible products as in the Swedish average, or $54 \cdot 10^9$ kJ.

A value for the energy content of wool is not available. One would need a better energy partition model for the sheep.

Value of hides ($5.9 \cdot 10^9$ kJ) is based on the Swedish average energy content per unit of weight and actual data.

In 1972, only pigs were exported live. The value for live animals ($7.5 \cdot 10^9$ kJ) is the energy content of the useful products that would have resulted if the pigs had been slaughtered on Gotland. It is not the total energy content of a live animal.

Actual energy content of manure is calculated to be $1900 \cdot 10^9$ kJ. (Fuel cost for producing equivalent amount of fertilizer is $289 \cdot 10^9$ kJ.)

Notes to Table 2.14

Indirect energy costs:[a-c]

Tractors = $4.09 \cdot 10^9$ kJ
Commercial fertilizer = $11.1 \cdot 10^9$ kJ
Plant protection = $0.25 \cdot 10^9$ kJ
Seed = $0.1 \cdot 10^9$ kJ
Machine rental = $6.3 \cdot 10^9$ kJ
Electrical energy used for mixing[a,c] = 2.9 MJ
Energy cost for producing oil crop grain[d] = 1.57 MJ
Energy contents of hides, meat, and live animals were calculated based on organic matter values; no value given to wool[e] = 31.52 MJ

Indirect energy costs associated with care of sheep[a,c,d]

Aluminum = $0.11 \cdot 10^9$ kJ
Medicine = $4.23 \cdot 10^9$ kJ
Advice and control = $0.07 \cdot 10^9$ kJ

[a] According to the Gotland County Agricultural Board and energy values for a sheep farm in Skepparp 6, Skåne, Sweden.
[b] Energy value of heating oil is approximated as diesel oil.
[c] Calculated according to Anonymous (1975), Johansson and Lönnroth (1975), Chapman (1974), energy use per tractor-hour, and assuming 40 hp per 500 female sheep.
[d] Anonymous (1975).
[e] Based on the total agricultural system on Gotland (raw data obtained from Farmek, Visby).

Notes to Table 2.18

Estimates of forest production are based on the fieldwork of Malkönen (1974) for similar conditions on Finland. This work shows different net productivities for different age classes. On Gotland, approximately 58% of trees are older than 71 years, 18% are less than 71 years, but more than 38 years, and the remaining are less than 38 years old (Anonymous 1972). The weighted net productivity for Gotland was about:

$$\text{Net prod. (trees)} = 0.58\ (6385\ \text{kg ha}^{-1}) + 0.18\ (5125\ \text{kg ha}^{-1}) + 0.24\ (3405\ \text{kg ha}^{-1})$$
$$\text{Net prod. (trees)} = 5443\ \text{kg ha}^{-1}\ \text{(dry weight)}.$$

Malkönen (1974) reported net productivity of understory vegetation at about 30% of tree production, or for Gotland 1633 kg ha^{-1}. Thus, the total net production is 7076 kg ha^{-1}, and gross production is 16,082 kg ha^{-1} (equivalent to $3 \cdot 10^8$ kJ ha^{-1}) or 1608 g m^{-2} yr^{-1}.

There are 1260 km^2 of economically productive forest land on Gotland, so that total gross production would be $38 \cdot 10^{12}$ kJ, of which $29 \cdot 10^{12}$ kJ is from trees and $9 \cdot 10^{12}$ kJ is from ground vegetation.

Forest area = 1260 km^2 and the average solar radiation = 10,184 kJ m^{-2} day^{-1}, so that total solar radiation = $4684 \cdot 10^{12}$ kJ yr^{-1}.

Estimated cuttings of timber for private use (Linné personal communication), 40,000 $m^3 \cdot (8.1 \cdot 10^6\ \text{kJ} \cdot m^{-3}) = 0.32 \cdot 10^{12}$ kJ, commercial use, 108,000 m^3 or $0.87 \cdot 10^{12}$ kJ.

Man-hours for cutting total = 234,000 (Anonymous 1972).

Export of raw wood products = 61,623 m^3 ($0.5 \cdot 10^{12}$ kJ). The total value of sales from lumber companies = $9.8 \cdot 10^6$ Skr. The estimated value of raw wood to timber industry = $6.25 \cdot 10^6$ Skr. Therefore, value of exports = $3.5 \cdot 10^6$ Skr.

Wages = $5.24 \cdot 10^6$ Skr; direct energy (oil + electricity) in oil equivalents = $24 \cdot 10^9$ kJ. Value of sales = $15.9 \cdot 10^6$ Skr. Value of raw materials = $6.3 \cdot 10^6$ Skr. Since fuel costs = $0.05 \cdot 10^6$ Skr, the value of wood = $6.25 \cdot 10^6$ Skr.

$$\text{Exported processed wood} = 7422 \cdot 10^3\ \text{kg} \cdot \frac{1\ m^3}{900\ \text{kg}} = 8247\ m^3$$
$$8247\ m^3 \cdot 8.1 \cdot 10^6\ \text{kJ}\ m^{-3} = 0.067 \cdot 10^{12}\ \text{kJ}.$$

From national accounts, $6.3 \cdot 10^6$ Skr of produced goods are exported and $9.7 \cdot 10^6$ Skr remains on Gotland.

Fuel costs for cutting wood, removing it from the forest, ditching, roads, and preparing the ground for planting were obtained for Sweden for small-scale forests (Genfors and Thyr 1976). These were expressed as cubic meters of wood harvested and kilometers of roads and ditches. The values used from Swedish forestries were:

Gasoline:	0.445 liters m^{-3} wood
Diesel:	1.38 liters m^{-3} wood
	9.4 m^3 km^{-1} for roads
	0.093 m^3 km^{-1} for ditching.

The gasoline required was calculated as 48.06 m^3, and diesel was calculated as 204 m^3. These represent an energy value of $8.76 \cdot 10^9$ kJ.

It was assumed that the energy costs of cutting for private individuals only included costs for cutting and removing:

$$\text{Gasoline: } 0.445 \text{ liters} \cdot m^{-3} \cdot 40{,}000 \; m^3 = 17.8 \; m^3$$
$$\text{Diesel: } 1.38 \text{ liters} \cdot m^{-3} \cdot 40{,}000 \; m^3 = 55.2 \; m^3.$$

This represents an energy value of $2.52 \cdot 10^9$ kJ.

The standing crop was based on average values obtained from measurements in south Finland (Malkönen 1974). For ground vegetation, this was calculated to be $1/3 \cdot (4450 + 5780 + 5490)$ kg ha^{-1} = 5240 kg m^{-2}. For the total forest system, the standing crop of ground vegetation is 524 g m^{-2} $(1260 \cdot 10^6 \; m^2)(18.5$ kJ $g^{-1}) = 12.2 \cdot 10^{12}$ kJ.

Based on the age distribution of trees (see above) and the south Finland study (Malkönen, 1974), the standing crop of trees is:

$$0.58\,(95{,}200) + 0.18\,(52{,}950) + 0.24\,(24{,}980) = 70{,}742 \text{ kg ha}^{-1}.$$

For the total forest area, the energy value of the tree standing crop is:

$$7074.2 \text{ g m}^{-2}\,(1260 \cdot 10^6 \text{ m}^2)(18.5 \text{ kJ g}^{-1}) = 164.9 \cdot 10^{12} \text{ kJ}.$$

References

Abramson, E. 1976. *Kosttabell (Diet Table)*. Ed. 5. Stockholm: Esselte Studium. 47 pp. (In Swedish).

Ackefors, H., and O. Lindahl. 1975. Investigations on phytoplankton production in the Baltic in 1973, 1974. *Medd. från Havsfiskelab. Lysekil No. 179*. Lysekil, Sweden: Marine Fisheries Laboratory (mimeo).

Ahlbom, H. 1982. Gotländsk Energi (Gotland's Energy). *Askö Laboratory*. University of Stockholm (mimeo, in Swedish).

Alm, G., and M. L. Nordberg, 1983. Accounting system for natural resources. Studies of changes of land use image analysis. *Remote Sensing Laboratory*. University of Stockholm.

Andréasson, I. M. 1984. Ekonomi och miljöpolitik. En regionalekonomisk studie av Gotland (Economy and environment: a regional economic study of Gotland.) *Department of Economics*. University of Stockholm (mimeo, in Swedish).

Andréasson, I. M., T. Hilding, and A. M. Jansson. 1983. Vatten Som begränsande faktor för ekologisk och ekonomisk produktion på Gotland (Water as a limiting factor for ecological and economic production on Gotland). *Nordiska Ministerrådet, Miljörapport*. pp. 240–249. (In Swedish).

Ankar, S., and R. Elmgren. 1976. The benthic macro- and meiofauna of the Askö-Landsort area (Northern Baltic proper). A Stratified Random Sampling Survey. *Contrib. Askö Lab. 11*. University of Stockholm.

Anonymous. 1945–72. Jordbruksstatistik. (Agricultural Statistics). *Official Statistics of Sweden*. Stockholm: Swedish Central Bureau of Statistics. (In Swedish).

———. 1972. Skogs-statistisk årsbok (Forestry—annual statistics). *Official Statistics of Sweden*. Stockholm: Swedish Central Bureau of Statistics (In Swedish).

———. 1975. Efficient use of energy. *Physics Today* 28:23–33.

———. 1977a. Vågenergi i Sverige (Wave energy in Sweden). *Nämnden för energiproduktionsforskning*. Stockholm. (In Swedish).

———. 1977b. Vindenergi i Sverige (Wind energy in Sweden). *Nämnden för energiproduktionsforskning*. Stockholm. (In Swedish).

———. 1979a. Förnybara energikällor, en sammanställning av aktuella bedömningar (Renewable energy sources, a review). *Liber Förlag*. Stockholm: Swedish Energy Research and Development Commission. (In Swedish).

———. 1979b. Elenergi—några siffror ur den senast tillgängliga statistiken (Electric energy—some numbers from the latest available statistics). *Vattenfall*. Stockholm (Brochure, in Swedish).

———. 1979c. Vindkraft till havs. (Wind power at sea). *Nämnden för energiproduktionsforskning*. Stockholm. (In Swedish).

———. 1981a. *Gotlands Energiverk AB*. Slite, Sweden: Personal communication.

———. 1981b. Hydrogeological map of Gotland County. *Sveriges Geologiska Undersökning*. Uppsala, Sweden.

———. 1982. *Vattenfall*. Stockholm: Personal communication.

Barnett, H. J. 1950. *Energy Uses and Supplies*. U. S. Bureau of Mines Information Circular 7852 (October).

Bennett, R. J., and R. J. Chorley. 1978. *Environmental Systems*. Princeton: Princeton University Press.

Berry, R. S., and M. Fels. 1973. Energy cost of automobiles. *Science and Public Affairs* (Dec. 29) 10 pp.

Boyden, S. 1979. An integrative ecological approach to the study of human settlements. *MAB Technical Notes 12*. Paris: UNESCO.

Brookfield, H. C., ed. 1980. Population-environment relations in tropical islands: the case of eastern Fiji. *MAB Technical Notes 13*. Paris: UNESCO.

Browder, J., C. Littlejohn, and D. Young. 1976. *The South Florida Study*. Center for Wetlands, University of Florida, Gainesville, Florida, and Bureau of Comprehensive Planning, Division of State Planning. Tallahassee, Florida.

Bullard, C. W. III, and R. A. Herendeen. 1977. The Energy costs of goods and services: an input-output analysis for the USA, 1963 and 1967. In *Energy Analysis*. J. A. G. Thomas, ed. pp. 71–81. England: IPC Science and Technology Press Ltd.

Byström, A., P. Clason, J. Linders, and C. G. Sundvall. 1980. Vågenergi för Gotland (Wave energy for Gotland). *Examensarbete inom gruppen för vågenergiforskning, CTH,* Göteborg, Sweden (mimeo, in Swedish).

Carlsson, D. 1979. *Kulturlandskapets Utveckling på Gotland.* En Studie av Jordbruks och Bebyggelseförändringar under Järnåldern (The Development of the Cultural Landscape on Gotland. A Study of Changes of Agriculture and Settlements During the Iron Age). *Liber Förlag,* Visby, Sweden. (In Swedish).

Chapman, P. F. 1974. Energy costs: A review of methods. *Energy Policy* 2:91–103.

Commoner, B. 1971. *The Closing Circle.* New York: Bantam Books.

Connell, J. H., and E. Orias. 1964. The ecological regulation of species diversity. *American Naturalist* 98:399–414.

Cook, E. 1971. The flow of energy in an industrial society. *Sci. American* 224 (3):135–144.

———. 1976. *Man, Energy, Society.* San Francisco: W. H. Freeman.

Costanza, R. 1980. Embodied energy and economic evaluation. *Science* 210:1219–1224.

Cottrell, F. 1955. *Energy and Society.* New York: McGraw-Hill.

Courant, R. 1936. *Differential and Integral Calculus.* Vol. 2. New York: Wiley Interscience.

Darmstadter, J., J. Dunkerley, and J. Alterman. 1977. *How Industrial Societies Use Energy (A Comparative Analysis).* Baltimore: John Hopkins University Press.

Eagleson, P. S. 1970. *Dynamic Hydrology.* New York: McGraw-Hill.

Ehrlich, P., and A. Ehrlich. 1970. *Population, Resources, Environment: Issues in Human Ecology.* San Francisco: W. H. Freeman.

Eneroth, B., K. Gustafsson, S. Persson, and G. Söderberg. 1979. Gotland 1990: Kommunal energiplanering för alternativ energiframtid (Gotland 1990: commune energy planning for alternative energy future). *Institutionen för Regional Planering, KTH.* Stockholm (mimeo, in Swedish).

Ersson, P. G. 1974. Kolonisation och ödeläggelse på Gotland (Colonization and devastation on Gotland). *Medd. Kult. Geogr. Institut.* University of Stockholm. (In Swedish).

Fairgrieve, J. 1921. *Geography and World Power.* New York: E. P. Dutton.

Gambel, A. B., 1964. *Energy R and D and National Progress,* Washington, D.C: Office of Science and Technology, Office of the President.

Genfors, W., and B. Thyr. 1976. *Basic Data for Energy Balance Evaluations in Forestry.* Garpenberg: Royal College of Forestry, Rep. No. 96.

Ghosh, A. 1958. Input-output approach in an allocation system. *Economica* Feb.:58–64.

Gilliland, M. W. 1975. Energy analysis and public policy. *Science* 189:1051–1056.

———, ed. 1978. *Energy Analysis: A New Public Policy Tool*. Boulder, CO: Westview Press.

Gotlands Kommun. 1974. *Visby: Staden Inom Murarna (Visby, the Town Within the Walls)*. Visby, Sweden: Gotlandstryck AB. (In Swedish).

Graneli, W. 1977. Sediment oxygen uptake in south Swedish lakes. *Inst. of Limnology*. Univ. of Lund (mimeo, in Swedish).

Hall, C., R. Kaufmann, and C. Cleveland. 1984. Time series analysis of energy and the U. S. economy. In *Proceedings of the Wallenberg Symposium "Integration of Economy and Ecology—an Outlook for the Eighties,"* A. M. Jansson, ed. Askö Laboratory, University of Stockholm. pp. 69–72.

Hannon, B. 1973. An energy standard of value. *Ann. Am. Acad. Polit. Soc. Sci.* 410:139–153.

Hesselman, H. 1908. Vegetationen och skogsväxten på Gotlands hällmarker (Vegetation and forest growth on heath land of Gotland). *Skogsvårdsföreningens Tidskrift* 6:63–167 (In Swedish).

Hilding, T. 1982. Gotlands Vatten, Erfarenheter av Miljöstatistik på Regional och Lokal Nivå (Gotlands Water, Experiences with Environmental Statistics at Regional and Local Levels). *Askölaboratoriet*. University of Stockholm (mimeo, in Swedish).

Holling, C. S. ed. 1978. *Adaptive Environmental Assessment and Management*. New York: John Wiley & Sons.

Hubbert, M. K. 1962. *Energy Resources*. Washington, D.C: Natl. Acad. Sciences. Natl Research Council.

Hytteborn, H. 1975. Deciduous woodland at Andersby, eastern Sweden. Above-ground tree and shrub production. *Acta Phytogeogr. Suecica* 61:1–100.

Intriligator, M. D. 1971. *Mathematical Optimization and Economic Theory*. Englewood Cliffs, NJ: Prentice-Hall.

Isard, W. 1972. *Ecologic-Economic Analysis for Regional Development*. New York: Free Press.

———. 1975. *Introduction to Regional Science*. Englewood Cliffs, NJ: Prentice-Hall.

Jansson, B.-O. 1972. Ecosystem approach to the Baltic problem. *Ecological Bulletins No. 16*.

———, and F. Wulff. 1977. Ecosystem analysis of a shallow sound in the northern Baltic. *Contributions Askö Lab* 18. University of Stockholm.

Jansson, A. M. 1985. Natural productivity and regional carrying capacity for human activities on the island of Gotland, Sweden. In *Economics of Ecosystem Management*, D. O. Hall, N. Myers, and N. S. Margaris, eds. pp. 85–91. Dordrecht, The Netherlands: Dr. W. Junk Publishers.

———, and N. Kautsky. 1977. Quantitative survey of hard bottom communities in a Baltic archipelago. In *Biology of Benthic Organisms*, B. F. Keegam, P. O. Ceidigh, and P. J. S. Boaden, eds. pp. 359–366. Pergamon Press: Elmsford, N.Y.

———, and J. Zucchetto. 1978a. Energy, economic and ecological relationships for Gotland Sweden, a regional systems study. *Ecological Bulletins No. 28.*

———, and J. Zucchetto, 1978b. Man, nature and energy flow on the island of Gotland, Sweden. *Ambio* 7(4):140–149.

———, and J. Zucchetto, 1980. A regional system study of the energy, economic and ecological aspects of the island of Gotland, Sweden. In *The Development and Application of Ecological Models in Urban and Regional Planning*. Man and the Biosphere Programme. National Committee of West Germany. UNESCO.

Johansson, T. B. and M. Lönnroth. 1975. Energianalyser—en introduktion (Energy analysis—an introduction). *Project Group for Energy and the Community*. Secretariat for Future Studies. Stockholm. (In Swedish).

Karlgren, L. 1972. Organic wastes from domestic sewage, agricultural activities and food processing. FAO/SIDA Training course on marine pollution in relation to protection of living resources. Göteborg, Sweden. May 2-June 3, 1972. SNV PM 220. *Swedish Environment Protection Board*. Stockholm.

Kelly, R. A., and W. O. Spofford, Jr. 1977. Application of an ecosystem model to water quality management: the Delaware Estuary. In *Ecosystem Modeling in Theory and Practice*, C. A. S. Hall, and J. W. Day, Jr., eds. pp. 419–443. New York: John Wiley & Sons.

Kemp, W. M., W. H. B. Smith, H. McKellar, M. Lehman, M. Homer, D. Young, and H. T. Odum. 1977. Energy cost-benefit analysis applied to power plants near Crystal River, Florida. In *Ecosystem Modeling in Theory and Practice*, C. A. S. Hall, and J. W. Day, Jr. New York: John Wiley and Sons.

Kjellström, B., and B. Gustafsson. 1976. Förstudie för långsiktig energiplan för Gotlands kommun. Seminarierapporter (Preparatory study for a long range energy plan for the municipality of Gotland. Seminar reports). *Inst of Thermic Energy Technology. Royal Institute of Technology*, Stockholm (mimeo, in Swedish).

Kylstra, C. 1974. Energy analysis as a common basis for optimally combining man's activities and nature. *The National Symposium on Corporate Social Policy*. Chicago.

———. 1981. Elda med ved och flis på Gotland. (Use of wood and wood chips for heating on Gotland). Visby, Sweden. (In Swedish).

Laszlo, E. 1972. *The Systems View of the World*. New York: George Braziller.

Leontieff, W. W. 1966. *Input-Output Economics*. New York: Oxford University Press.

Lieth, H. 1975. Modeling the primary productivity of the world. In *Primary Productivity of the Biosphere*, H. Lieth, and R. H. Whittaker, eds. pp. 237–264. New York: Springer-Verlag.

Limburg, K. E. 1980. Embodied energies of Sweden and their relation to the economy. *Department of Environmental Engineering Sciences*. University of Florida. Gainesville, Florida (mimeo).

———. 1983. Gotland's fisheries: a case study of the economic/ecological processes of renewable resource exploitation. *Askö Laboratory*, University of Stockholm (mimeo).

———, K. E., A. M. Jansson, and J. Zucchetto, 1982. A coastal ecosystem fisheries minimodel for the island of Gotland, Sweden. *Ecological Modelling* 17:271–295.

Lindahl, G., K. Wallström, and G. Brattberg. 1978. On nitrogen fixation in a coastal area of the northern Baltic. *Kieler Meeresforsch* 4:171–177.

Lindquist, S. O. 1980. *Director of Gotland's Fornsal*. Visby, Sweden: Personal communication.

Linné, I. *Regional Board of Forestry*. Visby, Sweden: Personal communication.

Ljungblom, L., H. Lundberg, A. Marklund, and S.-O. Sjöberg, 1978. Energiskog (Energy plantations). *Institutionen for kemisk teknologi, KTH*. Stockholm. (In Swedish).

Lönnroth, M., P. Steen, and T. B. Johansson. 1977. *Energy in Transition, A Report on Energy Policy and Future Options*. Stockholm: Secretariat for Future Studies.

Lotka, A. J. 1956. *Elements of Mathematical Biology*. New York: Dover Publications.

MacArthur, R. H., and E. O. Wilson. 1967. *Theory of Island Biogeography*. Princeton: Princeton University Press.

Magnusson, G. 1981. Biobränsle eller Kol (Biomass or coal). *Nämnden för energiproduktionsforskning*. Liber Förlag. Stockholm. (In Swedish).

Malkönen, E. 1974. Annual primary production and nutrient cycles in some Scots Pine stands communities. *Finnish Forest Research Inst.* 84(5):1–87.

May, R. M. 1973. *Stability and Complexity in Model Ecosystems*. Princeton: Princeton University Press.

Meadows, D., D. Meadows, J. Randers, and W. W. Behrens, III. 1972. *Limits to Growth*. New York: Universe Books.

Moberg, I. 1938. Gotland um das Jahr 1700. Eine kulturgeographische Kartenanalyse (Gotland A.D. 1700. A cultural geographical map analysis). *Geogr. Ann.* 1–2:1–112 (In German).

Munthe, H. 1913. Drag ur Gotlands odlingshistoria, sedd i relation till öns geologiska byggnad (Some aspects on the history of cultivation on Gotland, seen in relation to geological structure). *Sveriges Geologiska Undersökning*. Uppsala, Sweden. Ser. Ca: 11. (In Swedish).

Nijkamp, P., ed. 1976. *Environmental Economics*. Vols. 1 and 2. Leiden, The Netherlands: Martinus Nijhoff.

Nilsson, L. 1975. Midwinter distribution and numbers of Swedish Anatidae. *Ornis Scand.* 6:83–107.

Nilsson, T. 1982. Slutrapport från pilotprojektet Markanvändning/vattenkvalitet med Gotland som testområde (Final report from pilot project Land use/water quality with Gotland as a test case). *Dept. of Hydrology.* Uppsala University, Sweden (mimeo, in Swedish).

Nordberg, M. L. 1983. Markanvändning och markanvändningsförändringar i Lummelundaåns avrinningsområde, Gotland (Land use and land use changes in a drainage basin in Gotland). *Forskningsrapport 53,* ISSN 0346-7406. Naturgeografiska Institutionen. University of Stockholm.

Odum, E. P. 1971. *Fundamentals of Ecology.* ed. 3. Philadelphia: W. B. Saunders.

Odum, H. T. 1971. *Environment, Power and Society.* New York: Wiley-Interscience.

———. 1972. An energy circuit language for ecological and social systems: its physical basis. In *Systems Analysis and Simulation in Ecology,* B. C. Patten, ed. Vol. II. New York: Academic Press.

———. 1973. Energy, ecology and economics. *AMBIO* 2:220–227.

———. 1983. *Systems Ecology: An Introduction.* New York: Wiley-Interscience.

———, C. Kylstra, J. Alexander, N. Sipe, P. Lem, M. Brown, S. Brown, M. Kemp, M. Sell, W. Mitsch, E. De Bellevue, T. Ballantine, T. Fontaine, S. Bayley, J. Zucchetto, R. Costanza, G. Gardner, T. Dolan, A. March, W. Boynton, M. Gilliland, and D. Young. 1976. Net energy analysis of alternatives for the United States. In *Middle- and Long-Term Energy Policies and Alternatives: Volume I, Appendix. Hearings Before the Subcommittee on Energy and Power of the Committee on Interstate and Foreign Commerce, House of Representatives.* pp. 253–302. Washington, D.C.: G.P.O.

———, and E. C. Odum. 1976. 1981. *Energy Basis for Man and Nature.* New York: McGraw-Hill.

Olofsson, A. G. 1945. Gotlands Läns hushållningssällskap 1791–1941. Minnesskrift (County Agricultural Society of Gotland 1791–1941, memorial publication). *Almquist och Wiksell AB.* Uppsala, Sweden. (In Swedish).

Pattee, H. H., ed. 1973. *Hierarchy Theory: The Challenge of Complex Systems.* New York: Brazillier.

Patten, B., ed. 1971–76. *Systems Analysis and Simulation in Ecology.* Vols. 1–4. New York: Academic Press.

Persson, H. 1975. Deciduous woodland at Andersby, Eastern Sweden: Field layer and below-ground production. *Acta Phytogeogr. Suecica* 62: 1–76.

Pielou, E. C. 1975. *Ecological Diversity.* New York: Wiley-Interscience.

Pimentel, D., and M. Pimentel. 1979. *Food, Energy and Society.* Resource and Environmental Science Series, New York: Halsted Press.

Post, L. von. 1929. Om Gotlands myrar (Mires on Gotland). *Svenska Mosskultur fören. Tidskrift* 43:229–247 (In Swedish).

Richardson, H. W. 1972. *Input-Output and Regional Economics*. New York: John Wiley & Sons.

———. 1978. *Regional Economics*. Chicago: University of Illinois Press. IL.

Ricklefs, R. E. 1973. *Ecology*. Portland, Oregon: Chiron Press.

Rinne, I., T. Melvasalo, Å. Niemi, and L. Niemistö. 1978. Nitrogen fixation by bluegreen algae in the Baltic Sea. *Kieler Meeresforsch*. 4:178–187.

Rohde, H., R. Söderlund, and J. Ekstedt. 1980. Deposition of airborne pollutants on the Baltic. *Ambio* IX(3–4):168–173.

Shupe, J. W. 1982. Energy self-sufficiency for Hawaii. *Science* 216:1193–1199.

Sjöberg, S., and W. Wilmot. 1977. System analysis of a spring phytoplankton bloom in the Baltic. *Contr. No. 20, Askö Laboratory*, University of Stockholm.

Sjörs, H. 1954. Slåtterängar i Granggärde Finnmark (Meadows in Grangärde Finnmark, SW Dalarna, Sweden). *Acta Phytogeogr. Suecica* 34:1–135. (In Swedish).

Slesser, M. 1978. *Energy in the Economy*. New York: St. Martin's Press.

SMHI. 1979. Vattenföring i Sverige (Streamflow records of Sweden). *Swedish Meteorological and Hydrological Institute*. Gotab. Stockholm. (In Swedish).

SNV. 1969. Gotlands vattenförsörjning (Gotland's water supply). *Swedish Environmental Protection Board*. Stockholm. (In Swedish).

Soddy, F. 1935. *Wealth, Virtual Wealth and Debt*. New York: E. P. Dutton.

Söderberg, S. 1977. Sälar och sälskydd i Östersjöområdet (Seals and seal protection in the Baltic area). *Viltnytt* 6:6–12. (In Swedish).

Spencer, A. 1974. *Gotland*. London: David and Charles.

Spiller, G., A. M. Jansson, and J. Zucchetto. 1981. Modelling the effects of regional energy development on groundwater nitrate pollution in Gotland, Sweden. In *Energy and Ecological Modelling*, W. J. Mitsch, R. W. Bosserman, and J. M. Klopatek, eds. Amsterdam: Elsevier Scientific Publ. Co.

Stolt, B. O. 1971. Strandfågelfauna (Coastal birds). In *Gotlandskusten, översiktsplan för naturvård och friluftsliv (The Coast of Gotland—A Survey for Environmental Management and Open-air Recreation)*. National Board of Physical Planning and Building Report, Vol. 9: pp. 69–95. (In Swedish).

Svanberg, O. 1972. Händelsgodseln och vattnet. Litteraturreferat (Fertilizers and water. Literature survey). *Kungl. skogs-och Lantbruksakademiens tidskrift* 9 (Supplement):11–34. (In Swedish).

Sylwan, O. A. 1895. Martebo myr och dess utdikning (Martebo mire and its drainage). *Svenska Mosskulturföreningens tidskrift* 4:233–238.

Thomas, J. A. G., ed. 1977. *Energy Analysis*. Westview Press. Boulder, CO.

Van Dyne, G. M., ed. 1969. *The Ecosystem Concept in Natural Resource Management*. New York: Academic Press.

Viro, P. J. 1953. Loss of nutrient and the natural nutrient balance of the soil in Finland. *Commun. Inst. For. Fenn.* 42:1–51.

Von Bertalanffy, L. 1968. *General Systems Theory*. New York: Brazillier.

Wallentinus, H. G. 1973. Above-ground primary production of Juncetum gerardi on a Baltic sea-shore meadow. *Oikos* 24:200–218.

Watt, K. E. F. 1968. *Ecology and Resource Management*. New York: McGraw-Hill.

———. 1972. Man's efficient rush toward deadly dullness. *Natural History* 81(2):74–82.

———. 1973. *Principles of Environmental Science*. New York: McGraw-Hill.

———. 1974. *The Titanic Effect*. Stamford, Connecticut: Sinauer Associates, Inc.

———. 1982. *Understanding the Environment*. Newton, MA: Allyn and Bacon.

Whittaker, R. H., and G. E. Likens. 1973. Primary production: The biosphere and man. *Human Ecology* 1:357–369.

Wilson, A. G. 1981. *Geography and the Environment: Systems Analytical Methods*. New York: John Wiley, W. Sussex, U. K.

Young, D. 1975. Salt marshes and thermal additions at Crystal River, Florida. In *Power Plants and Estuaries at Crystal River, Florida*. H. T. Odum, et al., eds. pp. 281–372. Gainesville: Center for Wetland Resources, University of Florida.

Zucchetto, J. 1975a. *Energy Basis for Miami, Florida and Other Urban Systems*. Ph.D. dissertation. Gainesville: University of Florida. Department of Environmental Engineering Sciences.

———. 1975b. Energy-economic theory and mathematical models for combining the systems of man and nature, case study: the urban region of Miami, Florida. *Intl. J. Ecol. Mod.* 1:241–268.

———. 1981. Energy diversity of regional economies. In *Energy and Ecological Modelling*, W. J. Mitsch, R. W. Bosserman, and J. M. Klopatek, eds. Amsterdam: Elsevier Scientific Publ. Co.

———. 1985a. Relationship between total energy consumption and economic activity for the Florida economy, 1960–78. *Florida Scientist*. Florida Academy of Sciences, Orlando, FL 47(3):145–153.

———. 1985b. Energy-economic measures for selected economies of the world, 1960–81. In *Geography of Energy*. Dordrecht, Holland. D. Reidel Publishing Co. (In Press).

———, and A. M. Jansson. 1979. Total energy analysis of Gotland's agriculture: a northern temperate zone case study. *Agro-Ecosystems* 5:329–344.

———, A. M. Jansson, and K. Furugane. 1980. Optimization of economic and ecological resources for regional design. *Resource Management and Optimization*. 1–2:111–143.

———, and A. M. Jansson. 1981. Systems analysis of the present and future energy/economic developments on the island of Gotland, Sweden. In *Energy and Ecological Modelling*, W. J. Mitsch, R. W. Bosserman, and J. M. Klopatek, eds. Amsterdam: Elsevier Scientific Publ. Co.

Index

FSC
www.fsc.org
MIX
Papier aus verantwortungsvollen Quellen
Paper from responsible sources
FSC® C105338

If you have any concerns about our products,
you can contact us on
ProductSafety@springernature.com

In case Publisher is established outside the EU,
the EU authorized representative is:
Springer Nature Customer Service Center GmbH
Europaplatz 3, 69115 Heidelberg, Germany

Printed by Libri Plureos GmbH
in Hamburg, Germany